Realizing Tomorrow

Outward Odyssey
A People's History of Spaceflight

Series editor
Colin Burgess

REALIZING TOMORROW

The Path to Private Spaceflight

Chris Dubbs and Emeline Paat-Dahlstrom

FOREWORD BY CHARLES D. WALKER

UNIVERSITY OF NEBRASKA PRESS • LINCOLN AND LONDON

Library of Congress
Cataloging-in-Publication Data

Dubbs, Chris.
Realizing tomorrow: the path to private
spaceflight / Chris Dubbs and Emeline Paat-
Dahlstrom; foreword by Charles D. Walker.
p. cm. — (Outward odyssey: a people's history
of spaceflight)
Includes bibliographical references and index.
ISBN 978-0-8032-1610-5 (cloth: alk. paper)
1. Space flights. 2. Space tourism. 3. Outer
space—Civilian use. 4. Space industrialization.
I. Paat-Dahlstrom, Emeline. II. Title.
TL790.D79 2011
629.4'1—dc22   2010046898

Cover: Astronaut Dale A. Gardner, on an
extravehicular activity (EVA) during the flight
of STS-51A in 1984, holds up a "For Sale" sign,
referring to the two satellites that had just been
retrieved from orbit after their Payload Assist
Modules failed to fire. Astronaut Joseph P. Allen
IV is reflected in Gardner's helmet visor. Allen
would later become CEO of Space Industries International, Inc. that in the late 1980s attempted
to develop the Industrial Space Facility (ISF).
Shuttle-launched by NASA, the ISF would have
been the first privately financed and developed
man-tended space platform for commercial
research and manufacturing. Courtesy of NASA.

Set in Adobe Garamond.
Designed by R. W. Boeche and Kim Essman.

Contents

Illustrations

Following page 132

Foreword

It begins. From before recorded history, people like you and me have dreamed of journeying beyond the blue sky of Earth. We have dreamed of personal spaceflights even to other worlds. Within these pages you will discover the beginnings of those dreams coming true. This book is about the making of a new industry, new perspectives for humankind, and a new human movement.

My place in these beginnings came by accident—almost. During my pre-teen years in the 1950s, high-speed aviation, rockets, and space travel were constantly in the news. More than just headliners, the stories represented the beginnings of dreams come true—for a few. Chuck Yeager, Wernher von Braun, Willy Ley, and the Mercury astronauts were some of the people I associated with wishes I wanted to come true. They were real people doing and talking about real spacey things on TV and in the news.

While I imagined myself in the futures they were ushering in, there were other personages already there: Tintin, Buck Rogers, Dan Dare, Tom Corbett, Flash Gordon, Adam Strange, Tom Swift. I asked myself, could I be a part of these things? Would they become real and let me go along with them? My head was grounded just enough that I answered these questions with yes to the first and no to the second.

Within a few years, as the first cosmonaut and astronaut rocketed beyond the blue overhead, I focused on becoming a good aerospace engineer—good enough to build and launch great rockets and far-traveling spaceships. I didn't allow myself to consciously believe I could travel there, too. Flash Gordon I wasn't. Like so much else in life, that belief would change.

My thinking was changed by events. The revolution that hit the National Aeronautics and Space Administration, or NASA, in 1972 turned my head: delay human exploration of other worlds to instead build a reusable Space Transportation System. News reports said this would be a space shuttle to carry cargo and people into orbit and back to Earth; that meant astronaut pilots flying engineers and scientists. Working passengers going into space? Now, I could do that! But how?

That question, it turns out, was on the minds of many people. The dream was coming alive. Women and men in all walks of life were beginning to explore life paths intent on becoming nonprofessional space travelers. Any student of history could see that humankind's expansion into new realms continued as in the past. Explorers have always gone first into the unknown, but they are followed by the entrepreneurs, privateers, the gentry, and then the rest of us. As with terrestrial exploration and progress, adventuresome individuals would lead the way in the human movement into outer space.

In this book, Chris Dubbs and Emeline Paat-Dahlstrom relate some of the efforts to answer the "how" question through different visions of personal space travel. Many folks, including myself, watching the space scene had expansive and wonderful ideas for using space access for human benefit and in the process accomplishing the dream of personal space flight.

During the 1970s I became infatuated with the possibilities of space-based manufacturing and human settlements in space. Studying Gerard O'Neill's concepts for space colonies was a joy. I devoured all I could get about Krafft Ehricke's "extraterrestrial imperative." I melded with Harry Stine's "third industrial revolution." Public membership organizations, such as the grassroots L5 Society and the more staid National Space Institute (NSI), beckoned to me with their focus on fostering bigger commercial and government space programs. In 1977 I applied to be one of NASA's first space shuttle astronauts. L5 and NSI accepted me; NASA didn't. But I was already working on my backup plan.

Within the next decade my job at McDonnell Douglas Astronautics Company (MDAC) in St. Louis had connected me directly with space commercialization, space shuttle systems, and space flight operations. A good education, the focused pursuit of a goal, and some good fortune positioned me to take action. I became industry's first astronaut (and the first card-carrying L5 Society member making it into space).

The privately financed MDAC space commercialization project, Electrophoresis Operations in Space (EOS), was the first significant private investment in realistic space manufacturing. It paid my ticket into orbit. EOS succeeded technically but failed as a business. Earth-bound biotechnology process advances, compounded by the high cost and infrequent nature of space shuttle flights, killed our attempt at space industrialization. It was disappointing but pointed the way. Other applications of space and entrepreneurial means of accessing space were necessary.

I don't mean to suggest that NASA's actions in giving a U.S. senator, a U.S. congressman, laboratory researchers, internationals, and a teacher rides headed for space weren't important. Those actions were vitally important. Subsequent votes on Capitol Hill that advanced American space programs succeeded because of that senator's and that congressman's experiences. Investigations into rational pharmaceutical design, human immune, cardiovascular, neurological, and skeletal systems, and reactive materials science have all advanced thanks to the firsthand work in microgravity by expert scientists flying on NASA crews as payload specialist astronauts. And a teacher who lost her life, along with an industrial researcher and career NASA astronauts, onboard the *Challenger*'s final launch taught us much about ourselves and the need to strive. Christa McAuliffe also taught us to accept sacrifice in efforts to advance our understanding of ourselves and of the universe that surrounds us.

But progress has been too slow. Failures and successes in the pursuit of economic, safe, nongovernment space launchers unfold through accurate and personal accounts in the chapters that follow. I believe you, the reader, will see that entrepreneurial ingenuity should be applied more liberally to the development of space for societal and individual benefit. Government programs have a role in taking on the highest-risk pursuits, proving what is possible. The private sector can then exploit those proven paths while government takes on new challenges. The evolution of technology and its applications to meeting human needs and aspirations must be allowed and encouraged. My views and experiences parallel those of many of the individuals and organizations the authors have identified as touchstones on the visionary path to a space flight future.

In the last two decades of the twentieth century and more recently, I participated in the construction of space shuttle orbiters, microgravity research, and the planning and design of the International Space Station; led the National Space Society; counseled the X PRIZE; advised Space Adventures, Ltd.; and lectured at the International Space University, along with other activities, because I believe. I believe that when it comes to humanity's future, our expansion beyond our home planet is more than important and necessary—it's personal. It can happen only if both government organizations and individual ventures advance in that direction. Lead, follow, or get out of the way.

This challenge is not easy or risk-free. More fortunes will be lost, as will more lives. Soon thrilling jaunts touching the edge of outer space for a few minutes will become reality for growing numbers of rocket-propelled passengers. But accidents will happen. And we all must understand that as opportunities grow for personal space flight and for serious space research, so also will the technical difficulty, the energies involved, and the prospects of death. Suborbital flight is hard; orbital flight is much harder. Spaceflight in any form is not for the faint-hearted. But nothing really important is.

We, as individuals and organizations, must have ever-increasing opportunities to work and live in space and to make discoveries in the environments of spaceflight. Individuals need to see their living home planet as few ever have. We must create new technologies, new industries, and new wealth from an expanded human reality beyond Earth's blue skies.

Personal spaceflight is more than just a dream many share; it is a reality mankind must experience.

Chris Dubbs and Emeline Paat-Dahlstrom have collected the important efforts to date in pursuit of that dream. They spotlight the trends and the leaders to watch, to follow, and to join as our dreams become real. Read on, get excited, find involvement.

Charles D. Walker
Industry's first astronaut
Space shuttle missions 41-D, 51-D, and 61-B

Acknowledgments

For all of the people who shared our excitement about this subject and gave generously of their time in so many ways, we wish to say thank you. For all of those who shared memories, documents, and photographs, we enjoyed spending time with you and hope that this book captures your many stories.

In particular, we want to acknowledge the help provided by the following:

Ken and Gretchen Davidian
Don Davis
Peter Diamandis
Jeff Foust
Mike Gould
Gary Hudson
Richard Johnson
Ryan Kobrick
Jeff Manber
Gregg Maryniak
Doug Messier
Anita Miller
John Moltzan

John Oelerich
Robert Pearlman
Carol Perry
Fell Peters
Dan Slater
Space Exploration Technologies
(SpaceX)
Dean Truax
Scott Truax
Virgin Galactic
Per Wimmer
X PRIZE Foundation

Special thanks to Eric Dahlstrom, whose interviewing, writing, and editing skills, along with his extensive knowledge about space, played an important role in the creation of this book.

We also want to acknowledge the considerable role played by Colin Burgess, series editor for the Outward Odyssey series. His editorial guidance was invaluable in steering us through the writing and rewriting of this book.

Realizing Tomorrow

Prologue

Due to lack of interest,
tomorrow has been canceled.

Richard Hoagland,
following the launch of *Apollo 17*,
the final manned lunar mission

The Holland America cruise ship ss *Statendam* stood at berth in New York Harbor on 4 December 1972 preparing for a curious mission related to the American space program. Tom Buckley, reporter for the *New York Times*, boarded the ship, unsure what to expect. There was a buzz that this trip would be something special, with big-name headliners: Wernher von Braun, head of the American space program; Apollo astronaut Edgar Mitchell, who eleven months earlier had walked on the moon; and writer Arthur C. Clarke, whose novel *2001: A Space Odyssey* had been made into a movie in 1968.

Space artists Don Davis and Rick Sternbach boarded the ship about that same time, toting portfolios of their artwork. They were ambitious youngsters of nineteen and twenty-one attending science fiction conferences when they had been recruited for the cruise. For a little help with graphic design and audiovisual work, they got free passage and a chance to display their art. Sternbach had already drawn pictures of the *Statendam* flying between the Earth and the moon for the promotional brochure.

At 2:00 p.m., science fiction writer Isaac Asimov walked up the ship's boarding ramp with some trepidation. He had a mild case of agoraphobia and would have been more than content to stay put in Manhattan. He had traveled on ships twice before, and neither of the experiences had been pleasant. But the chance to be part of the *Statendam* cruise proved too strong a lure.

Asimov would be joined onboard by other science fiction writers, Robert Heinlein, Frederik Pohl, Ben Bova, and Theodore Sturgeon. A host of luminary scientists rounded out the passenger list: Carl Sagan, NASA adviser and director of Cornell's Laboratory for Planetary Studies; German-born rocket designer and space visionary Krafft Ehricke; Marvin Minsky, the man breaking ground with artificial intelligence at the Massachusetts Institute of Technology (MIT); radio astronomer Frank Drake, director of the Arecibo Observatory, who was pioneering the search for extraterrestrial intelligence; and physicist Robert Enzmann, who had developed ideas about nuclear-powered rockets.

As Asimov would later explain in one of his regular articles in *The Magazine of Fantasy and Science Fiction*, he had been recruited the previous spring by Richard Hoagland, "who was aflame to lead a party of idealists" to see the launch of *Apollo 17*. Hoagland's vision was to gather scientists, artists, astronauts, visionaries, science fiction writers — the best space minds — to watch the final Apollo moon mission, then engage in seminars to discuss man's future in space. The advertising brochures labeled it the "Voyage Beyond Apollo," an apt name, since this was to be the final Apollo moon landing, the end of an era.

If on the day he boarded the *Statendam* Asimov had carried in his pocket a copy of the previous day's *New York Times*, he could have read von Braun's tribute-lament to the Apollo program: "With the end of the Apollo lunar program now upon us, one of the most important chapters in space exploration ends. Greater feats of exploration of the planets lie before us."

The Apollo program had been seen as a stepping stone to the "greater feats" America planned for space. When the first Apollo spacecraft touched down on the lunar surface in 1969, nine additional manned moon landings were planned, Apollo missions 12 through 20. That same year, the government Space Task Group proposed in its report, *America's Next Decades in Space*, an ambitious plan for an Earth-orbiting space station by 1980, a lunar-orbiting station for fifty to one hundred people, a reusable shuttle, and the first Mars mission in 1983. The cost of all this would run in the neighborhood of $4 to $8 billion a year, depending on how aggressively it was pursued.

The problem was that NASA's budget had been dropping like a spent rocket booster since 1966. President Richard Nixon had referred to the triumph of the *Apollo 11* moon mission as "the greatest week in the history of the world since the Creation." Subsequent flights, however, began to appear like reruns of an old TV show, the same grand achievement over and over. Apollo began to lose much of its magical hold on politicians and the public.

Funding cuts in the Apollo budget first eliminated mission 20 and delayed missions 18 and 19. Then 18 and 19 also fell victim to the shrinking budget. The flight of *Apollo 17* would bring a close to this glorious chapter in space exploration. The time was right for taking stock, for some serious soul searching and redirecting to gain a new vision of America's future in space.

The second day out, the serious business of the cruise began, with lectures and panel discussions that continued for nearly a week. All three headliners—von Braun, Clarke, and Mitchell—failed to appear, but the show went on without them. Asimov got things rolling with a lecture about colonizing space by building cities inside hollowed-out asteroids.

Asimov had written about an asteroid belt inhabited by pirates in his 1953 juvenile novel, *Lucky Starr and the Pirates of the Asteroids*. But now he spoke about mining and colonizing asteroids as though this possibility waited just over the horizon. This was the sort of grand idea that the audience wanted to hear, the sort of bold project that could top a moon landing.

Next came Norman Mailer, who had a literary connection to the Apollo program. *Life* magazine had sent him to chronicle the history-making *Apollo 11* launch, which resulted in the book *Of a Fire on the Moon*. He had been impressed by the buttoned-down, corporate efficiency of NASA, but troubled by it as well. He had come to that project looking for romantics and bold adventurers, pioneering the new frontier of space. He found instead immensely competent technicians and self-effacing astronaut heroes; "robots and saints, adventurers and cogs in the wheel" was how he put it in the book. That same ambiguity still haunted him on this occasion. He believed in the necessity for humans to voyage into space, because it was "part of our human design, part of our inner imperative," and yet he also saw man's exploration of space as a challenge to God. Instead of merely gathering rocks on the moon, NASA should be conducting experiments in extrasensory perception, levitation, and magic.

Carl Sagan reported on the latest photographs of Mars. Just a year earlier, *Mariner 9* had become the first spacecraft to orbit another planet and had revealed an exciting vista of craters, riverbeds, massive extinct volcanoes, and canyons. The signs of erosion seen in the photographs offered the first evidence that liquid water may have flowed on the planet. Sagan calculated that if all the frozen ice and carbon dioxide on Mars melted, it would release enough carbon dioxide into the atmosphere to create a greenhouse effect and warm the planet.

Krafft Ehricke might have been excused a sigh of impatience as he sat through Sagan's lecture. All those Mars views were very interesting, but he had already devised a project that would take humans to Mars and warm the planet. In the late 1950s, while employed at General Dynamics, Ehricke had worked with a team that planned out an eight-man, 450-day Mars mission. It would be an orbital mission, with an option for a small, piloted lander that supported two people for seven days on the Martian surface. The project timeline called for a Martian Excursion Vehicle lander to be tested on the moon in November 1972.

Like von Braun, Ehricke was both a visionary and a popularizer. His rocket roots ran deep. One of the Peenemünde scientists who came to the United States after the war to jump start the U.S. space program, he had already done distinguished work for NASA and private space firms. He liked large ideas and aggressive timelines—nuclear-powered rockets, space hotels, space hospitals, space tourism, colonies on Mars.

At General Dynamics, he had proposed a gargantuan rocket named Nexus. Nearly as tall as the Washington Monument, 164 feet wide, and with a launch weight of some twenty-four thousand tons—just about the weight of the *Statendam*—Nexus would have been capable of delivering nine hundred tons of payload to orbit.

Don Davis sat in the audience while Krafft Ehricke lectured about using giant orbiting mirrors to increase agricultural production on Earth and generate nighttime solar power. The same mirrors could be used around Mars to warm small areas of the planet and make them more habitable. Like most space visionaries, Ehricke saw space as a frontier to be developed for the benefit of mankind, and he was a strong proponent of space industrialization. On the scale being considered, industrial production and mining activities in space and on the moon and asteroids could eventually send thousands of technicians into space.

Although only nineteen years old, Davis was perhaps more familiar with the topography of the moon than any of the Apollo astronauts. As a high school student in 1968, he had marched into the offices of the U.S. Geological Survey's branch of Astrogeologic Studies carrying a painting he had done of the moon and was hired on the spot. They would utilize his considerable talents on their Lunar Earth Side Geologic Map project, a nearly decade-long endeavor to illustrate the geologic evolution of the moon.

"Being young and brash, I cornered Ehricke with a lot of questions after his lecture," Davis recalls. "Then he invited me back to his cabin to continue our discussion." Ehricke talked about his orbital mirrors idea, but also reminisced about fighting on the Russian front during the war, before joining von Braun at Peenemünde.

"He told me about seeing the first v-2 success and then being in the gathering that evening in which Walter Dornberger announced the beginning of the Space Age. That was the last thing we spoke about before he shooed me off to bed." Davis was touched by Ehricke's generosity in sharing his time. "He seemed to be a genuinely sincere visionary, with a sense of a mission in life which shone behind his features and accent. I think the daring nature of his grand schemes was a kind of conceptual precursor to later bold ideas like space colonization and space elevators."

Excitement built all day on 6 December in anticipation of Apollo's first night launch. The cruise ship had anchored seven miles off the Florida coast, yet the giant Saturn rocket was clearly visible on its launchpad. That evening the deck filled with excited passengers. Through binoculars, Don Davis studied the rocket on its illuminated launchpad. In the opposite direction, to the east, silent lightning indicated the passing of a distant storm. Someone's transistor radio crackled out music but offered no coverage of the countdown.

Asimov found a good vantage point on deck. He had never seen a rocket launch, a curious omission for a man whom some considered the best at spinning tales about space. At thirty seconds to liftoff, a hold came in the countdown that would eventually extend for two hours and forty minutes, pushing the launch to 12:33 a.m. the next day.

When it finally occurred, Asimov was immediately taken by the silence. The rocket seemed to levitate silently from the ground atop a brilliant plume of flame that lit the sky from horizon to horizon under a copperish dome of light. The sound arrived forty seconds later as a rumbling earthquake that enveloped the ship and reverberated in the hull.

For space artist Rick Sternbach, the launch was all about visual images and color, "the repeating shockwaves off rocket, the blowtorch yellow-orange glow around the vehicle, the smoke and steam streaming away in every direction." He had witnessed the daytime launches of *Apollo 11* and *Apollo 13*, but this was an altogether different experience.

After the launch, the ship's many bars filled with celebration and discussion. Ehricke estimated to a gathered crowd that the brightness of the night launch was about that of five hundred full moons. "Incomparably beautiful," Robert Heinlein termed it. For Norman Mailer, "It was the one time when I wanted instant replay." Eighty-two-year-old novelist Katherine Anne Porter, on assignment to cover the launch for *Playboy* magazine, never expected to witness anything like it in her life. "I came out of a world so primitive you can scarcely imagine it," she said. "We barely had gaslight in New Orleans when I was a girl. When I saw them take off, I wanted with all my soul to be going with them."

Fresh from his own rounds of celebration, Richard Hoagland commandeered the ship's public address system to announce that "due to a lack of interest, tomorrow has been canceled" — as though the launch were so singular an event that all else lost meaning in its wake. The comment might have served as a final epitaph for the extraordinary Apollo program, except that the *Statendam* passengers had gathered precisely to consider "tomorrow" and how to fill its possibilities.

That evening they gathered for a cinematic glimpse of tomorrow as portrayed in the movie *2001: A Space Odyssey*. The "space" part of the film began with the *Orion III* spaceplane cruising with ballet-like precision to a docking with a wheel-shaped space station in orbit between Earth and the moon.

Director Stanley Kubrick had taken great pains to portray spaceflight realistically. He had drawn on the expertise of aerospace companies and tapped the imagination of technical artist Harry Lange, who had worked on advanced designs for NASA, which explained why *Orion III* shared the same conceptual design as NASA's space shuttle, announced earlier that year. However, unlike a NASA spacecraft, *Orion III* was a commercial vehicle, emblazoned with the logo of Pan American Airlines, one of the largest air carriers at that time.

Curiously, ever since the film's release in 1968, the same year that the *Apollo 8* mission orbited the moon, Pan Am had been receiving phone calls from people who wanted to be on the first commercial flight to the moon. Given that the technology was obviously now in place to take people to this destination, they assumed that commercial flights couldn't be far behind.

After Neil Armstrong and Buzz Aldrin actually landed on the moon in July 1969, the number of registered moon-flight hopefuls soared to twenty-five thousand. That same month the travel magazine *Holiday* featured the moon as its destination of the month. By the time the *Statendam* passengers viewed the launch of the final Apollo mission, Pan Am's moon list, now called the First Moon Flights Club, included the names of more than eighty thousand worldwide travelers eager for a lunar excursion. Pan Am estimated that if the cost was calculated by its then current rate schedule of 6 cents per mile, the moon trip would cost $14,000 one-way.

For the *Statendam* passengers, *2001* resonated with total feasibility and gave visual form to the near-term space future. After all, a space shuttle and a giant ringed space station had been part of the explore-space playbook drawn up by von Braun in the 1950s. In newspaper and magazine articles and TV specials, von Braun had proposed a wheel-shaped station, 250 feet in diameter, rotating to maintain gravity, with an onboard staff of eighty people. Supplied by cargo-carrying spacecraft like the space shuttle, the station would become a staging site for excursions to the moon and the planets. Within a generation, scientists and their families could be living on the moon and welcoming tourists.

With NASA's shuttle now on the drawing board, it seemed as though this vision would soon become reality. By routinely shuttling back and forth to space, the shuttle would "help transform the space frontier of the 1970's into familiar territory," as President Nixon described it when announcing the program.

Less than thirty-five years had elapsed between Charles Lindbergh's solo flight across the forbidding Atlantic Ocean and men setting foot on the moon. For more than half that time, visionaries such as von Braun and Krafft Ehricke had touted the feasibility and inevitability of man's thrust into space. Along with popular writers and filmmakers, they had planted the idea that despite the obvious dangers and limitations of space travel, space was a viable, even desirable, place to be.

The Apollo program demonstrated that we had the technology to deliver on that vision. We were now poised to transform space from the exclusive preserve of government programs and highly trained test pilots into a more accessible frontier of science, commerce, and habitation. In one lifetime, private companies would begin to harvest the bountiful resources of space, and private citizens would be visiting and living in that new frontier.

1. The Entire Population of the Earth in Orbit

We now have the technological ability
to set up large human communities in space.

Gerard K. O'Neill,
The High Frontier, 1976

A few weeks before the launch of *Apollo 17,* a physics class at Hampshire College in Amherst, Massachusetts, hosted a guest speaker lecturing on the feasibility of creating mammoth orbital space colonies. Not space *stations* of the sort then on NASA's drawing board but space *colonies* on which tens of thousands of people, eventually billions of people, would take up permanent residence. It was a concept more akin to science fiction than science fact, something beyond even the futuristic speculations of Wernher von Braun or Krafft Ehricke.

And yet the speaker who faced this class of skeptical undergraduates was an unassuming Princeton physics professor named Gerard K. O'Neill. O'Neill came with impeccable credentials, having established his reputation in the mid-1950s by publishing an influential article on how particle accelerators could capture and store electrons. He worked with a team at Stanford's linear accelerator to build the first storage ring, now basic to all high-energy particle accelerators.

In 1966 O'Neill took a leave of absence from his Princeton teaching to join NASA as one of their new scientist astronauts. Although chosen as part of a group of sixty-eight candidates to undergo testing at Brooks Air Force Base, he was not selected as an astronaut.

By 1972 O'Neill's prodigious intellect and imagination had put aside the notion of colliding subatomic particles or flying in space and had been totally captured by the even grander idea of building orbiting space colonies.

It had come to him innocently enough three years earlier, through a class-room exercise in an advanced section of his Physics 103 class. The year was 1969, and the students had been following the progress of the Apollo program. With this in mind, O'Neill had the class use the principles of physics to examine the problems of having people and things in orbit.

The topic question of their final paper was, "Is a planetary surface the right place for an expanding technological civilization?" The unanimous conclusion was a resounding no. For any number of reasons, they concluded, the surface of the Earth—or the surface of any planet—was not the best environment for human life. In fact, it might well be the worst! Planets were a creation of nature, too random and accidental to be the best habitat for humans: too much useless mass, inaccessible resources, the whims of weather.

Certainly, in this high-technology, spacefaring age, if a bunch of geniuses put their minds to it, they could come up with something far better. They could *design* a world for maximum efficiency, specifically for human habitation. Instead of putting the population on the planetary surface, the "outside" of the planet, the students reasoned, why not put it "inside" the planet instead?

The exercise had been designed to engage the students, which it certainly had. But it had a more profound effect on their teacher. The concept, the reality, of orbital space colonies carried enough technological challenge to drive an obsession. The shape, size, land area, orbital location, atmosphere, material requirements, gravitation, shielding, sunlight, food production, mining and manufacturing in space—there was plenty for a nimble brain to wrestle with.

Over the next few years, as O'Neill slowly formulated his ideas about space colonies, he became convinced that it was an important issue that represented not only an alternative vision for the American space program but for humanity as well. He slowly mustered the necessary scientific and engineering underpinnings for the concept. How much material would you need to construct a colony? Where would it come from? How would you engineer it to withstand the structural forces? How would you harness solar energy? The issues were many, the problems mammoth. But concepts emerged, theories evolved, mathematical formulas got committed to paper. O'Neill wanted to lay them before the scientific community to see if they passed muster.

In 1974 he organized a conference on space settlement at Princeton University. The modest turnout included a few academics, a few students, and astronaut Joe Allen. Allen was one of NASA's new scientist astronauts. O'Neill had met Allen in the mid-1960s when he himself had tried unsuccessfully to become an astronaut. The presence of Allen, who was also a nuclear physicist, gave added credibility to the conference.

Plans were to begin small, the conference participants concluded, with a colony large enough for two thousand intrepid space settlers. It would be located at an orbital position relative to the Earth and moon called a Lagrange point, named for the eighteenth-century French mathematician Joseph Louis Lagrange. At Lagrange point 5 the colony would be equidistant from the Earth and the moon, where the gravitation of both bodies balanced, and the colony could stay put without expending any fuel to do so. These pioneers would take building materials from the moon and asteroids and shoot them into space with an electromagnetic accelerator. Using power harnessed from the sun by solar power satellites, they would forge that raw material into the building blocks of their habitat.

The colony would exist inside a can-shaped cylinder, 100 x 200 meters, rotating slowly so that centrifugal force simulated 1 g, or full Earth gravity. Mirrors would reflect sunlight into the interior. All of this might take fifteen or twenty years. Once this prototype was established, larger colonies would be built for ten thousand people and separate ones for space manufacturing.

It was a bold and visionary plan, rescued from being science fiction by the credentials of the participants and the scientific rigor they brought to the examination of the issue. "By the middle of the next century most 'dirty' industries could be operating off the Earth, using nonpolluting technology," O'Neill commented at the time. Indeed, as the participants pointed out, in a world threatened by nuclear holocaust and catastrophic pollution, space colonies provided insurance for the continuity of the human race. Eventually, most of humanity might be resident in space colonies and the Earth would be a "worldwide park," a nice place to visit for a vacation.

All of this might have escaped much notice if not for the presence at the conference of Walter Sullivan, science reporter for the *New York Times*. Good space news was hard to come by in 1974. The final Apollo missions had been canceled, and NASA's budget was in freefall. Here was a guy who wanted to send thousands of people to live in space.

Sullivan's article about the conference ended up on the front page of the *Times* and launched a media feeding frenzy. The BBC, CBC, and a host of radio and television stations called O'Neill for interviews. The phone rarely went silent. The *Los Angeles Times*, *Washington Post*, and *Time* magazine all did articles. And months later, when O'Neill's scholarly article about space colonies appeared in the journals *Physics Today* and *Nature*, it put a further stamp of scientific respectability on the concept.

The publication of the *Physics Today* article was one of those moments that happen in science when a single event, a single idea, captures a pent-up force. The article may have been one of the most photocopied and circulated in history, getting passed along to colleagues and friends in the academic and scientific communities. People called O'Neill and wrote to him with objections, suggestions, or questions, which forced him to refine the technical details and calculate specifically how things would work.

"I called him up and asked how he planned to pay for his space colony," remembers Mark Hopkins. Now the senior vice president of the National Space Society, in 1974 Hopkins was an economics grad student at Harvard. He had cofounded the Harvard-Radcliffe Committee for a Space Economy, and the group had been looking for some large project that could lay the groundwork for making space affordable. In response to Hopkins's question, O'Neill mentioned the use of solar power satellites of the sort conceived by Peter Glaser, who pioneered the concept of massive arrays of solar collectors in geosynchronous orbit. These large collectors in space would convert solar energy to microwaves and transmit it to Earth for use as electricity.

Space artist Don Davis read about the Princeton conference in a *Palo Alto Times* article titled "Space Colony in Our Times." Thirty-three years later he still had the article, a creased and yellowed memorial to the gravity of that long-ago moment.

Space artists had a well-established role in promoting space ideas to the public. They gave form to the frontier of space. What did it mean to travel in space? What did the planets look like? In the 1950s, when von Braun wrote a series of articles on our space future for the popular *Collier's* magazine, the accompanying illustrations, by legendary space artist Chesley Bonestell, captivated readers more than von Braun's words.

Davis, a great admirer of Bonestell's work, had first shown his own paintings to the master in 1969 and since then had paid occasional visits to Bonestell's home in Carmel, California. "He was ruthless and also enlightening to show one's work to. He told you what he thought and you listened. I came away from every visit determined to do better the next time."

Davis immediately phoned Gerard O'Neill, explaining that he wanted to paint a perspective view of the interior of a space colony and that any details would be helpful. O'Neill sent him some descriptions.

Keith and Carolyn Henson, two individuals who would play a considerable role in popularizing O'Neill's ideas, read the *Physics Today* article and also contacted O'Neill. They were engineering students at the University of Arizona, and they wanted to be involved. Carolyn Meinel Henson came by her space interest genetically, being the daughter of two famous astronomers, Aden and Marjorie Meinel. Keith, on the other hand, took the more prosaic route of a lifetime interest in science fiction. They expressed their ideas about agriculture on a space colony and were invited to present at the next conference, planned to take place at Princeton the following year.

Meanwhile, NASA was also showing interest in O'Neill's ideas. It provided a small grant to underwrite the cost of a second conference and for O'Neill to hire as his assistant a newly minted Princeton PhD in physics, Eric Hannah. NASA also agreed to conduct a summer study session on space colonies with Stanford University at its Ames Research Center, Moffett Field, California, the following year.

At NASA's Johnson Space Center (JSC), Hubert "Hugh" Davis headed up the Advance Systems Office, tasked with thinking up space projects of the future. He had helped develop the propulsion systems on the Apollo lunar lander, which by definition made him a rocket scientist. JSC decided to send him to the second Princeton conference to see what O'Neill was up to.

The second Princeton conference, held in May 1975, provided the first directed study of the thorny issues related to O'Neill's concepts. About twenty-five invited speakers from academia, business, and NASA presented papers.

Also in attendance were experts in agriculture and magnetism, launch vehicles, and solar power satellites, as well as Joe Allen and students from MIT and Harvard who had got wind of the magnitude of the undertaking and wanted to be involved.

O'Neill now talked in terms of a Model 1 or Island One colony, housing ten thousand people. This was to be the prototype; once constructed, it would become the base for building the next, larger colony, and so on. "If you look at the growth rates that you could get from that first one," O'Neill predicted in a contemporary interview, "then you'd probably be talking about a quarter of a million people by the year 2000. Because you'll be going up very fast after you get the first beachhead."

The conference was a whir of activity: presented papers and ad hoc groups grappled with such heady ideas as growing food crops in space and shielding the colony from radiation. This was mind-boggling stuff that energized everyone. "I have never been to a conference where there was more energy and excitement," Mark Hopkins recalled.

Hugh Davis, who presented a paper on the heavy lift vehicles needed to boost materials to low Earth orbit and to the colony's final orbital destination, got the impression from the conference that O'Neill was an absolute master at inspiring young people, getting them to work nonstop at developing these ideas.

Among the key points they established was that the quantity of material that would have to be brought up from Earth to begin construction was well within the capabilities of 1970s technology and could be delivered by the space shuttle, then in development.

Since it was now known that aluminum, titanium, iron, and oxygen were abundant on the moon, those minerals would account for more than 95 percent of the materials needed for construction. They would be mined on the lunar surface, then launched into space via a magnetic catapult called a mass driver, to be retrieved and used by the colonists.

Colony agriculture also got considerable attention, including in the paper delivered by Carolyn and Keith Henson. Grains, fruits, and vegetables would do well in the colony, as would goats, chickens, and rabbits, making the colony self-sustaining in regard to food.

The grander vision now emerging for the initial space colony was of a massive cylinder 3,300 feet long with its axis pointed toward the sun. Alternating strips of land and windowed areas ran the length of the cylinder. Outside mirrors would reflect sunlight into the colony, and a parabolic reflector would focus twenty-four hours of solar energy onto a power generator. Slow rotation would simulate Earth's gravity.

What truly required a conceptual leap for the average person trying to grasp the O'Neillian vision was that life in this habitat would be very Earth-like. No cramped space capsule or sterile metal container but wide-open space, cozy homes, a lush landscape, babbling brooks. Don Davis had brought along his first series of paintings portraying the interior of Model 1, and they showed an idyllic environment of green hills, streams, and lakes, with separate residential and agricultural areas—in fact, a life very much like what the colonists would be leaving behind on Earth.

Cost guesstimates for the construction of Colony One ran approximately $100 billion, about four times the price tag for the Apollo program. Mark Hopkins gathered estimates for the various project components from the participants. If those estimates were accurate, he concluded, then the colony could actually pay for itself by exporting power back to Earth from solar power satellites. This possibility of economic viability delighted O'Neill, since it represented another piece of the puzzle falling in place.

O'Neill proposed that the construction of space colonies should become a national commitment. The colonies were not some long-term dream but feasible, self-sufficient habitats achievable by the end of the century. With their solar power satellites, the colonies were a better approach to energy independence than nuclear power plants and their hazardous stockpiles of plutonium.

Not everyone shared O'Neill's enthusiasm for the idea. In the report he made back at JSC, Hugh Davis highlighted the educational value of the conference and characterized space colonies as a far-in-the-future thing. "The way I saw it, he had 10,000 people living in a huge habitat that's built with materials launched from the moon with a device that didn't exist." Davis didn't think there was anything NASA could do with O'Neill other than pay attention to his ideas and support the good work he was doing inspiring students.

Despite Davis's sober assessment of space colonies, NASA had already planned a summer study session on space settlements at Ames Research

Center in California. Summer sessions were a tradition at Ames. Cosponsored by the American Society for Engineering Education, these programs were offered for academics as something of a crash course in systems engineering. In one five-week session participants had to design all facets of a complex project.

About two dozen participants attended the NASA-Ames/Stanford University Summer Study on Space Colonization in 1975, academics mostly, plus the students from the Princeton conference and the Hensons, along with speakers from NASA. Richard Johnson from Ames and Bill Verplanck from Stanford served as codirectors. In preparation for the session, Johnson had contacted artist Rick Guidice to do paintings of the colonies. "I told him to put stuff in there that looked spacey," Johnson recalled.

But Don Davis continued his direct link to O'Neill, who was now telling him to imagine the view of San Francisco Bay from Sausalito—vast space, verdant hills. So Don painted that scene, complete with the Golden Gate Bridge, and hurried the painting down to Ames, literally with the paint still wet.

The session got off to a fast start. Verplanck, an engineer, had a background in team building and came with a supply of 3 x 5 cards, felt pens, and tape. They set up parameters. The colony had to be a safe place to live for ten thousand inhabitants and provide for all of their physiological needs. Colonists had to be able to obtain a supply of raw materials and process them and needed access to a reliable transportation system, and the colony had to possess a thread of economic viability. Those rules locked things in place.

Ideas were written on the cards and taped to the wall. Someone else came along and taped another card with a modified idea. It took all of three hours, and everyone was assigned to a discipline team with a specific assignment. Clearly, this was not your usual academic conference.

O'Neill wasn't on hand for the start of the session and came and went throughout, but he was taken aback by the direction of things. By the time he appeared, the group had already made some fundamental decisions that he didn't like.

The problem was safety, according to Johnson, gravity and shielding. The colony would have to spin to simulate gravity. The only thing anyone knew for sure about long-term survival in gravity was normal Earth gravity, or 1 g, so that was the calculation they put in. O'Neill had calculated

1 g by having his tube rotate at 3 rpm. But the Ames participants determined that wouldn't work and settled instead for a redesigned colony that only had to rotate at 1 rpm to achieve 1 g.

Protecting the colony's inhabitants from cosmic radiation presented an even greater challenge. The group had discounted the possibility of some form of magnetic shielding since no one knew how to build such a thing, opting instead for the known quantity of passive shielding. This involved surrounding the colony with as much mass as possible to protect it from the ionized particles constantly zipping through space. The only material available for constructing such a shield was lunar regolith, rocky material from the moon. About six feet of rocky shielding surrounding the colony would do the trick nicely. Unfortunately, that would require a breathtaking ten million tons of rocky debris, all of which had to be mined on the moon and lifted to the colony.

If that wasn't troublesome enough, they determined that this shield could not actually be attached to the colony, because that much mass rotating at the same speed as the colony would literally tear the thing apart. The solution would be to have the shield separate from the colony, with a colony that did spin and a shield that did not.

O'Neill didn't like these conclusions, as Johnson recalled. These were problems that O'Neill thought would kill the whole thing. "We had a minor revolution," as O'Neill worked with some of the students and professors to come up with alternative solutions. But they couldn't make it more plausible.

"O'Neill would say, 'I'll go away and recalculate this and make it work.' He was always saying something like that," according to Johnson. "But we had given him a pretty bitter pill to swallow."

"It was a dream," is how Johnson sums it up. "The thing with O'Neill was that he kept on saying that it was not a dream. That was his real hangup; he couldn't let it be something that might one day happen; he wanted it to happen in his lifetime."

However, in the final analysis, such technical challenges did more to strengthen O'Neill's case than weaken it. If one type of shielding had to replace another type of shielding, if that would take more time and more money, then so be it. The space colony project had been a work in progress since 1969 and had slowly evolved as more experts weighed in with their criticisms and suggestions and as the continual process of study refined the design.

O'Neill had already come to the firm conclusion that space colonies were well within the range of present-day technology. There were plenty of challenging technical issues to deal with, but building a space colony was similar to any other large civil engineering project, such as the Alaskan pipeline or the Panama Canal. The only truly critical element in the equation was the public will and commitment to support the idea.

The Summer Study on Space Colonization ultimately provided a detailed summary of its conclusions titled *Space Settlements: A Design Study* (1976). Never before had the subject received this level of scrutiny, from the physical properties of space and the human needs of colonists to details on the design, construction, costs, and operation of the colony. Some problems remained, but so detailed an exploration contributed to the gathering credibility of the idea.

Nor did the writers of the report express any doubt as to the weightier implications of what was being proposed. Once begun, the migration of the human species to settlements in space would have a profound effect on human history, redefining how we viewed our home planet and our future. The conclusion of the report put it this way:

Since a considerable portion of humanity — even most of it — with ecologically needed animals and plants — may be living outside the Earth, the meaning, the purpose, and the patterns of life on Earth will also be considerably altered. The Earth might be regarded as a historical museum, a biological preserve, a place which contains harsh climate and uncontrolled weather for those who love physical adventure, or a primitive and primeval place for tourism. This cultural transition may be comparable to the transitions in the biological evolution when the aquatic ancestors of mammals moved onto land. . . . The opportunity for human expansion into space is offered; it needs only to be grasped.

In July 1975, midway through the Ames Summer Session, O'Neill testified before the U.S. House of Representatives Subcommittee on Space Science and Applications of the Committee on Science and Technology. By this point he had become an adept promoter of his ideas.

O'Neill assured the congressmen that this was not science fiction but well within "right now" technology. The space shuttle, then under development, would be able to support construction of the colony. There might be a $100 billion price tag on the first colony, but solar power satellites, constructed in conjunction with the colony, would not only lead to energy independence but eventually make the United States an energy exporter.

This was sweet music to congressmen who had oversight of NASA's budget. The Apollo program had cost nearly $30 billion. Sure, they had beaten the Russians to the moon, but what had they got for their investment besides a bunch of moon rocks? Here was a space project promising a financial return. Not only would thousands of constituents get to go into space, but it could potentially free the United States from dependence on Middle East oil, an addiction that had recently become dramatically more expensive following the creation of the Organization of Petroleum Exporting Countries (OPEC).

The publication of *Space Settlements: A Design Study* proved to be a turning point in the space colony phenomenon. It was an exhaustive examination of a "space habitat where 10,000 people work, raise families, and live out normal human lives." It also placed the imprimatur of NASA on the project. The Summer Session was a NASA program, at a NASA facility, and the conference findings were compiled and published by NASA. This gave the whole concept incalculable credibility.

In truth, NASA had no interest in building space colonies. They were on a scale far beyond NASA's means and far beyond anything it could sell to Congress. However, NASA was interested in identifying any of the colony technologies that might apply to its own projects.

In the report, the shape of the colony had undergone a fundamental change. O'Neill's original straight tube was now a torus, the Stanford Torus, a tube 427 feet in cross section, wrapped into a wheel that was one mile in diameter. This was the classic bicycle tire colony design of the movie *2001* and of von Braun's imagination, the sort portrayed in the popular media for decades.

Living and agricultural areas would occupy the ring-shaped tube of the colony, which would be connected by six spokes, forty-eight feet in diameter, to a central hub where incoming spacecraft docked. Rotation of 1 rpm simulated actual Earth gravity in the outer ring. A large stationary mirror above the hub illuminated much of the interior of the colony with reflected sunlight.

In fact, sunlight was a critical resource for the colony. Solar energy would provide ample electricity as well as power solar furnaces to refine metals from ores launched from the moon. These metals would be used to construct solar power satellites and additional colonies.

During the years of its greatest popularity, O'Neill's colony designs utilized three basic shapes: sphere, torus, and cylinder, in Islands One, Two, and Three. A spherical space colony, O'Neill's Model 1, was a design first proposed in 1929 by John Desmond Bernal. Bernal speculated that a burgeoning population on Earth would eventually have to move off the planet. He conceived of numerous self-sufficient, orbiting spheres, each ten miles in diameter, each home to up to thirty thousand people.

Spheres had the advantage of being better able to hold pressure and to minimize the necessary amount of shielding. O'Neill adopted a more modest design of about one mile in diameter, rotating at 1.9 rpm to achieve Earth-normal gravity. Agricultural production would be separate from the living area, in stacked torus rings arranged beneath the sphere.

The largest colony design, Model 3, and the one advocated by O'Neill in his landmark book, *The High Frontier: Human Colonies in Space* (2000 [1977]), was composed of two large cylinders. The cylinder shape offered the maximum habitable space. Linking the cylinders stabilized their orientation toward the sun, which would allow exterior mirrors to illuminate the interior through long strips of windows. Each cylinder stretched for 18 miles and had a diameter of 1.8 miles. Combined, they offered habitation for ten million people.

If a "perfect storm" is the convergence of several weather patterns that create an event greater than the sum of their individual effects, then such was the case with the launch of L5, the space colony advocacy group born from the 1975 Princeton conference. Its creation blended the growing scientific interest in space settlement with the pent-up public interest in space that had reached its peak with the moon landings. Then it catalyzed the brew with an activism that ignited a populist movement.

O'Neill's call for the colonization of space represented more than just another space program option. He wanted an optimistic future for mankind, he would often say. The suggestion, of course, was that mankind's future was not so optimistic, a theory trumpeted by a spate of popular doomsday books published in the late 1960s and early 1970s. Titles such as *The Population Bomb* (1968), *The Late Great Planet Earth* (1970), and *The Limits to Growth* (1972) prophesied overpopulation, uncontrolled pollution, resource depletion, and economic collapse. Their philosophy, known collectively as limits to growth (LTG), offered a bleak view of human future.

In Krafft Ehricke's visionary future there were no limits to growth because there were no limits to humans' creativity. Science and technology would deliver us from these ills. O'Neill shared this view. If humans had thoughtlessly imperiled their home planet, then scientists would replace old worlds with new ones. If people had exhausted Earth's resources, then space colonists would tap the limitless power of the sun and the boundless materials on the moon and in asteroids. Huge clusters of colonies would eventually occupy Lagrange points 4 and 5, with the capacity for billions of people. Eventually there would be more people living in orbit than on the surface of Earth.

The limits-to-growth philosophy became a way of thinking, a cultural mindset or meme that influenced the way popular culture defined a set of challenges facing the world. Space colonization arose as a counter-meme that offered a more optimistic future many people were ready to embrace. History appeared to be at a turning point where humanity had a choice between a finite planet with a static population rationing ever-dwindling natural resources or expansive space settlements with unlimited resources and the opportunity to create new civilizations. It was an idea whose time had come.

Keith and Carolyn Henson were cut from a different cloth than the academic scientists and NASA engineers who delivered papers in Princeton. They were entrepreneurs and science fiction fans who had protested the Vietnam War in college and still maintained a counterculture edge. Unlike most of the others in attendance at the Princeton conference, the Hensons actually wanted to go into space and live in a colony. While other participants were content to study the technical issues involved in creating a colony, write papers, and present at conferences, the Hensons wanted to make it happen.

Following the Ames Summer Session, the Hensons sent a solicitation to approximately three hundred people on a list of names O'Neill had compiled, inviting them to join an organization that would advocate for O'Neill's ideas. About thirty people joined, and the L5 Society was born. "L5" stood for Lagrange point 5, the orbital location for the first colony. The society began very much as a shoestring operation in a back room of the business the Hensons ran, called Analog Precision.

The first issue of the group's newsletter in September 1975 announced their mission as educating the public about the benefits of space communities and manufacturing. The group would serve as a clearinghouse for news and information and would raise funds to support work on these concepts. Their long-range goal was to disband the society in a mass meeting at Lagrange point 5, or L5. "L5 by '95" became the slogan for their goal of having an O'Neill colony in operation within twenty years.

That first issue also included a letter of support for space colonies from Arizona congressman and contender for the presidency Morris Udall. Udall was a liberal Democrat, but the standard line about L5 members was that they were 5 percent Democrats, 5 percent Republicans, and 90 percent anarchists. Early L5 members skewed toward the idealistic, but the group would eventually attract such divergent members as arch-conservative Senator Barry Goldwater and psychedelic drug advocate Timothy Leary.

Whether they were technological utopians looking to build a commune in the sky or pragmatic engineers, they all found a banner in L5. However, because L5 early on attracted such rabid let's-go-to-space-now members who carried something of a science fiction air, O'Neill kept a careful distance between himself and the group.

Operated largely by the Hensons and an odd assortment of volunteers, or "groupies," as Keith Henson termed them, L5 quickly achieved its goal of being a clearinghouse for space settlement issues and activities. In the first year, the *L5 News* published reports from Washington, conference announcements, research developments, space colony activities at various universities, articles by NASA officials, reports of O'Neill's activities, technical articles, book reviews, and letters from supporters.

Aerospace engineer Greg Bennett first heard of the L5 Society when he attended the World Science Fiction Convention in 1976. The speaker had a slide presentation and was talking about creating giant colonies in space, huge things with tens of thousands of people living in them. There were financial break-even models and an analysis of solar power satellites. He seemed completely serious.

But the thing that really struck Bennett, the thing that would stay in his mind for years, was the illustrations of the people living in this colony. They showed images of "people picnicking on beautiful hillsides overlooking cozy townships . . . real people doing real-people things in bucolic woodland settings," Bennett would recall many years later. "This was a far cry from the sterile presentation of a handful of astronauts circling Earth in a metal marvel."

This was the image connecting with many people, what made O'Neill's ideas resonate with the public—they represented a paradigm shift in the way we thought about space. Space travel would not be just a tiny cadre of test pilots in the confines of a capsule but ordinary people living in space. The rationale for our presence in space had made a giant shift, away from exploration and toward self-sustaining permanent settlement. It was a mind-blower.

The presenter Bennett heard on that occasion was Keith Henson. Armed with the technical details from Princeton and Ames and with slides of the Guidice and Davis paintings, Henson had taken the space colony vision on the road. Nor was he alone. Bennett had soon formed the first L5 Society chapter and was himself making similar presentations, by his count exactly one hundred of them over the next two years.

Eric Hannah, O'Neill's assistant, was doing the same thing, spending much of his time speaking at universities, museums, and community groups, filling the auditorium at the Jet Propulsion Laboratory in Pasadena, California, and fielding questions from Rotarians in Indiana. Hannah still recalls one of the questions he got from a high school science teacher about life in a space colony: "What would you do with the dead bodies?" Hannah didn't tell the guy that Keith Henson, the resident agricultural expert, had already suggested that they would make good compost.

The message continued to draw the attention of the media as well. In July 1976 *National Geographic* published a special Bicentennial issue that included a piece by Isaac Asimov titled "The Next Frontier?"—a fictionalized account of a visit to an O'Neillian space station. In glorious *National*

Geographic detail, the article showed illustrations of space colonies under construction; colonists at shopping malls; vast, glass-enclosed agricultural areas; mining operations on the moon, with a mass driver; and a giant solar power station ten kilometers across to beam power to the colony and back to Earth. "Eventually," according to Asimov, "space colonies and power satellites may be as plentiful as milkweed in the wind, and Earth's energy crisis a forgotten episode."

Although a fictionalized account, the editors pointed out that the article was not science fiction but a visualization of the outcome of a serious study that had been conducted the previous year (the Ames Summer Session), the results of which would be published in August. Dr. Richard Johnson of Ames was listed as the contact for additional questions.

"Over the course of the next six to seven years, I got about 10,000 letters of response," Johnson recalled in a 2007 interview. "They represented a Rorschach of responses: How can we help? How can I volunteer? Who do I contact?" He answered every single letter and sent each a copy of the final report. "It was that article that pushed things to the limit. Things snowballed after that."

NASA didn't know quite what to do with space colonies. In the face of a continually shrinking budget, it seemed unreasonable to give them serious attention. Federal funding had begun to shift to social programs and finding solutions to today's problems. "The public is 'now' oriented," NASA administrator James C. Fletcher told a November 1975 conference of the National Academy of Engineering.

"The average person pays lip service to the kind of world he wants for his grandchildren, but he is really interested in what affects him now," Fletcher summed it up. He might bear all of the costs associated with a meteorological satellite not because of his interest in space activities but because of the immediate benefit of a better weather forecast. But "selling him on the idea of financing a $100 billion space colony for the 1990s must be viewed by him in the same context as convincing an Eskimo that he needs a refrigerator for his igloo."

Despite the hurdles to long-range planning, Fletcher said that we should not be deterred from embracing a vision of the future. To Fletcher's way of thinking, space offered humanity an "alternative future." We could squander that potential and become another of history's failed experiments, or "we can accept the challenge of the great spaces between the worlds and establish our citadels among the stars."

This was unusually florid rhetoric for a NASA administrator. In part, it came as a reaction to the report of the House Committee on Science and Technology, before which O'Neill had testified that summer. Among the committee's recommendations was that NASA focus on an overarching concept that "should represent one or more mind-expanding endeavors which would challenge the imagination and capability of the country"—in other words, something grand enough to top the Apollo moon landings. At the same time, the committee thought that such a program should provide a substantial return and immediate benefit for the investment the United States continued to put into its space program. The "now" orientation once again reared its head.

If that charge neatly defined the quandary NASA faced in a post-Apollo world, it also mapped the strategy for O'Neill and the L5 Society. It was one thing to captivate a science fiction buff with the notion of living in space or to engage an engineer with the challenge of a lunar mining operation; it was something altogether different to get funding to make it happen.

O'Neill sometimes used the term "ignition point" to refer to a moment in time when all of the pieces fall into place and a project can finally unfold. Although he never shied away from speaking of the grander philosophical and historical significance of space colonies, in the near term his focus, and that of L5, would have to be defined by the incremental developments that moved the project inevitably to that decisive moment.

Among the elements critical to progress on space colonies were (1) an operational space shuttle that could deliver payloads to space frequently and inexpensively, (2) a solar power satellite that could provide power to the colony and beam it back to Earth, (3) a mass driver to fling raw materials from the moon into space, and (4) proof of the feasibility and economic viability of industrial-scale manufacturing in space.

Fortunately, there was a considerable "now" quotient in these elements. The space shuttle, then undergoing operational testing, would substantially reduce the cost of lifting payloads into space, and with its large payload capacity and frequent launch schedule it would support a wide range of scientific, military, and commercial space projects.

Peter Glaser's concept of transmitting power via microwaves made solar power satellites theoretically possible, as well. Glaser worked closely with O'Neill during these years. Given the concern over the depletion of fossil fuels and the OPEC oil embargo of 1973, energy independence was about as *now* as one could get.

Space industrialization became the focus of subsequent Summer Sessions at Ames in 1976 and 1977. The July 1975 joint U.S.-Soviet Apollo-Soyuz mission had already conducted experiments in the separation of chemicals in weightlessness. Space manufacturing and materials processing could be a multibillion-dollar business by the 1980s, especially in such areas as pharmaceuticals, crystals, tungsten carbide components, turbine blades, and many more. G. Harry Stine had already published his book, *The Third Industrial Revolution* (1975), about the dawning of the space industrial age. He claimed humans were passing from the age of space exploration to the age of space exploitation.

The coming revolution moved toward critical mass during 1976 in the pages of the *L5 News*. In the March issue, NASA's director of advanced programs, Jesco von Puttkamer, informed readers that government attitudes toward space had changed dramatically in the past year. A new wind was blowing in Washington, he asserted. The idea of space colonization was no longer incredible to politicians. The first space facility, a virtual construction camp, might be ready by 1983–84, with space colonization beginning in the late 1980s or early 1990s.

Puttkamer had already issued contracts to Grumman Aerospace and McDonnell Douglas to design a space station, referred to as a "construction shack" for its role as a base of operations for the construction of solar power satellites and other manufacturing operations. The station, which would be home to two hundred people, would be constructed in low Earth orbit out of materials brought up on the space shuttle, then would be boosted to geosynchronous orbit by solar electric propulsion.

That same March 1976 issue carried an interview with Puttkamer's deputy, Robert Frietag, who shared the agency's vision on other space projects near and dear to the heart of L5 members, including the development of nuclear energy for use in space; servicing satellites in space; building large structures in space for many purposes; mining the moon or the asteroids; reflecting sunlight from huge mirrors to illuminate Earth; putting hospitals in space, particularly to treat conditions like heart trouble; and establishing a base for disposing of nuclear wastes. "We could even have, in years to come, a Hilton Hotel in space."

All in all, everything seemed on track in 1976. *L5 News* reported the slow buildup of interest in space colonization and the fledgling grassroots activism of L5 and its members to engage the public and decision makers on this issue. The Ames Summer Sessions continued to drive interest. The report from the 1975 session was published in 1976, and a second Ames session further studied space manufacturing.

Readers of the January 1977 *L5 News* might have overlooked a small article of large portent, "Space Base Put on Ice." Hidden behind articles on space as an evolutionary imperative and a dual interview with Robert Frietag and counterculture guru Timothy Leary, the article hinted at how outside influences would soon affect the best-laid plans for space colonization and change the way L5 did business. It reported that funding for NASA's proposed Manned Orbital Facility, regarded by some as the "construction shack" for space colonies, had been removed from the 1978 budget.

Gathering momentum behind space colonization promised to turn space colonization into a genuine cultural movement. Two books would serve as handbooks for enthusiasts, O'Neill's *The High Frontier* and T. A. Heppenhiemer's *Colonies in Space*. At Princeton, O'Neill founded the Space Studies Institute to promote research about space settlements. And, just coincidentally, the first *Star Wars* movie drew record crowds. Membership in L5 climbed into the thousands.

Something truly unprecedented had begun to jell around the colony movement. Not only had engineers and scientists embraced colonies as a suitable vehicle for applying their disciplines, so too had academics in the humanities and social sciences. People who had never before been involved with or shown an interest in space, who had taken only passing notice of the moon landings, now saw a connection to what they were doing and began to write scholarly papers about it. Anthropologists, sociologists, historians, psychologists, architects, political scientists, lawyers, and economists lent their voices to the issue of space settlement. An entire academic underpinning began to grow from this interest: conferences, workshops, undergraduate and graduate courses, and journal articles.

The dark cloud on the horizon, the problem hinted at by that small article about canceled funding, was political policy. The administration of U.S. president Jimmy Carter planned to reduce spending on space and refocus on social issues.

L5 reported to its members the budgetary battles in Congress as two NASA projects fought for survival, solar power satellites and the Jupiter Orbital Probe (JOP). Incensed by the removal of JOP from the budget, the American Astronomical Society sprang to action. Its members phoned and telegraphed Congress and spoke to the local media. They set up a telephone tree, each member calling seven people and asking them to call seven more to inform them about JOP. It proved effective. In the end, JOP remained in the budget, while a project dear to the heart of L5 members, SPS, got cut. The project that was to have weaned the United States off foreign oil, that was to have been the economic foundation for space colonies, lost funding. It was a bitter lesson for L5.

The "new wind blowing in Washington" that Jesco Puttkamer had spoken of just the previous year as heralding growing support for space settlements had shifted direction. When the TV news program *60 Minutes* aired a segment on space colonization in October, it reported on reader response the following week, including a letter from the fiscal watchdog in the U.S. Senate, William Proxmire. "It's the best argument yet for chopping NASA's funding to the bone," he stated. Famous for his Golden Fleece awards, which he bestowed on wasteful government spending projects, Proxmire's disapproval of a project drew attention. "As Chairman of the Senate Subcommittee responsible for NASA's appropriations, I say not a penny for this nutty fantasy."

Not surprisingly, the *L5 News* carried the publication's first article on political lobbying that same month. Others would follow. Clearly, it wasn't enough to raise public awareness about colonies or to excite members with scientific breakthroughs. To get movement on the many issues relating to space colonization required federal funding. That meant swaying the views of politicians, and they responded to different issues and tactics. L5 was about to change course.

The 1980s began with a stunning lobbying success for L5 on an issue that was bedrock belief for the group — private enterprise in space. The United

Nations had drafted an international treaty, the Agreement Governing the Activities of States on the Moon and Other Celestial Bodies, that recognized all the objects in space as the common heritage of mankind. This meant that these bodies and any resources derived from them belonged to all nations.

Fundamental to the O'Neill scenario of space colonization was the mining of resources from the moon and near-Earth asteroids with which to construct colonies. It would not do to have such operations regulated by the United Nations, potentially obstructed by member nations, and to pay dividends to countries that made no investment and assumed no risk in the enterprise. The effect of the "Moon Treaty" would be to discourage private investment and limit the freedom of space colonists.

"On the Fourth of July 1979 the space colonists went to war with the United Nations of Earth." Keith Henson stated his hyperbolic opposition to the treaty in the January 1980 issue of the *L5 News*. "The Treaty makes about as much sense as fish setting the conditions under which amphibians could colonize the land."

The L5 Society hired a Washington lobbyist, who trained key L5 members to circulate through the halls of Congress explaining opposition to the treaty. The society launched a publicity and letter-writing campaign, and in short order Mark Hopkins threw together a telephone tree to inundate Congress with calls. When the United States ultimately refused to sign on to the treaty, L5 had scored a major success. People started to take notice of this upstart organization flexing its political muscle.

Hopkins likes to tell the story about what happened shortly after the defeat of the Moon Treaty. Two FBI agents showed up at the modest L5 headquarters in Tucson and wanted to know why L5 was sending copies of its newsletter to the Soviet embassy. The poor intimidated clerk went to check the records and came back to report that they were sending the newsletter because the embassy had paid its dues and was a member of L5.

It was later revealed that the FBI was investigating L5 because they had learned that the KGB was investigating L5. As it turned out, the United States had been one of the strongest supporters of the Moon Treaty in the United Nations, and the Soviet Union had been attempting to block it. The KGB was trying to figure out why the L5 Society could stop the treaty when they couldn't. It was one small measure of the group's growing political influence.

The Moon Treaty success clearly set a new path for L5. While the group kept the same long-term goal of space colonization, it moved away from a single focus on O'Neill's ideas. To some extent, members became even more engaged during these years as the society's lobbying efforts required frequent requests for their participation. At the same time, L5 made a hard turn from its initial attitude toward achieving space settlements in twenty years to a general pro-space advocacy. An O'Neillian colony became a distant goal.

Michael Michaud, in his insightful book on the pro-space movements of the 1970s and 1980s, *Reaching for the High Frontier,* pointed out that O'Neill, and by extension the L5 Society, experienced the full evolutionary cycle of a pro-space movement: (1) enthusiasm for the big idea, (2) encounter with the realities of government and politics, leading to frustration, and (3) the scaling down of near-term goals.

Hinting at the tensions within the organization, one of L5's directors, Eric Drexler, wrote an article for the October 1983 newsletter, reassuring members that the society's focus on the general development of space remained consistent with the group's early goals. But there were many paths to those goals, he assured them, and practical projects had to pave the way to the grand visions.

By the mid-1980s, as preparations got under way to merge the energetic, activist L5 Society with the more sedate National Space Institute, established in 1974 by Wernher von Braun, Keith Henson penned his own lament to L5 and the O'Neillian vision:

[L5] has, sadly, been much less successful in accomplishing goals implicit in the meme, or even — if we are honest — in making noticeable progress in that direction. In 1975, the founders were expecting a program (such as SPS) to start by the early 80s. Space colonies would follow naturally from large scale economic activities, and we hoped to disband in space by the Society's 20th anniversary. Here, in the tenth anniversary year, some issues of this magazine go by without a mention of space colonies or projects that might lead to them, and we are about to merge with NSI. . . .

Memes as replicating structures can die out or become completely inactive. Most memes lose their intense hold on people with the passage of time, especially when the promise of the meme is at great variance with reality. So it is with the space colony meme. The gradual displacement of human habitation with a general pro-space theme in the Society and the pending merger with NSI and loss of a clear goal are by-products of this divergence.

It is as fine an epitaph for O'Neill's vision of space colonies as any. In 1987 the L5 Society officially merged with the National Space Institute, and the combined groups took the name National Space Society, which continues to this day, advocating for a strong space program.

The first grand vision to seed thousands of people in space within a generation had lost steam. It was as though the country had for a decade been caught up in a dream of an alternative destiny and now awoke to wonder what had gone wrong. And yet the movement launched by O'Neill and promoted by the L5 Society had fired the public imagination with the idea that space was not only for astronauts but for private citizens as well.

The hard work of actually getting them to space would become the task of innovative government programs and a few inspired individuals.

2. The Birth of Private Rocket Companies

I am not prepared to say whether it was the
promise of taking mankind into a completely new
realm or merely the earthshaking roar that I found so
fascinating. In any case, working eight hours a day on
rockets was never for me quite enough.

Robert Truax

In 1965 auto racing legend Walt Arfons showed up at the Sacramento, California, headquarters of the rocket company Aerojet-General with a simple problem. He needed more power in his rocket car so that he could recapture the land speed record he and his brother Art had previously held. During 1964–65 the land speed record was as ephemeral as a mayfly. Replacing the internal combustion engine with a jet engine had allowed drivers to set nine new records in those years. In a running duel between the Arfonses and Craig Breedlove, speeds got pushed from 434 mph to over 600 mph.

Walt had taken the next logical step by creating a rocket car, using the Jet Assisted Take Off (JATO) rockets produced at Aerojet. He kept adding more and more of the standard JATOs to increase the speed. His current Wingfoot Express II utilized twenty-five of the small rockets. Unfortunately, it was a short, fast ride. Each JATO produced one thousand pounds of thrust for about twelve seconds, not long enough to maintain top speed through the timing traps.

Arfons met with Robert Truax, a rocket engineer working at Aerojet after an illustrious career in the U.S. Navy, where he had helped to develop the JATO. Truax headed Aerojet's Advanced Development Division, a position that allowed him to apply his somewhat contrarian rocket ideas to designing truly massive, low-tech rockets launched from water that would lift huge payloads into orbit at a fraction of what it cost NASA.

Arfons's dilemma intrigued Truax. He briefly considered mating Arfons's car to a twenty-thousand-pound-thrust Corporal missile, but couldn't free one up from government inventory. Instead, they turned their attention away from the land speed record and concentrated on Arfons's real bread and butter, the drag racing circuit. Arfons owned some jet engine–powered drag racers that he ran as special attractions at drag strips. He was always interested in having a faster car with which to thrill the crowds.

Truax had developed a steam-powered rocket on his own and thought that it might be just what Arfons needed to tear up the track. How fast could it take him down the quarter mile, Arfons wanted to know. "How many g's can you take?" Truax responded. "If you can take ten, I can get you across the finish line in 2.87 seconds, and you will be going 626 mph." That was just the sort of give-me-fame-or-give-me-death talk that Walt Arfons liked to hear. He hired Truax to build him a steam-powered rocket engine for his dragster.

Along with associates Bill Sprow, Facundo Campoy, and Ed Rice, Truax formed Truax Engineering, Inc. to build the engine. Working in Truax's Saratoga, California, garage and Campoy's nearby welding shop, they converted a surplus Air Force oxygen tank into a rocket by fitting it with a nozzle and a plug operated by a hydraulic actuator.

There was beauty in the simplicity of the design, and the more experience Truax gained with rockets, the more he appreciated simplicity. The rocket consisted of a tank, a heating unit run on propane, and a throttle. Tap water served as the rocket's reaction fluid — heated and pressurized until released as steam, like air rushing from a balloon. With the water heated to 475°F, under 500 psi pressure, the rocket could produce 4,700 pounds of thrust. Arfons set the practical limits lower, however, 417°F at 300 psi, because that combination would propel the car to 3.5 g, which was considered the limit the driver could endure without wearing a jet pilot pressure suit.

When Traux Engineering finished the engine and Arfons had fitted it into his dragster, he arranged a trial run at the Akron, Ohio, airport so that the media could see this land rocket in action. With only a modest two hundred pounds of pressure in the tank, veteran driver Bobby Tatroe slipped into the cockpit. Although the engine was capable of being throttled, like a true drag racer Tatroe slammed the accelerator to the floor. Steam blasted out the rear of the dragster, shooting the car from 0 to 260 mph in four seconds.

It was at this point that a mild cross wind and a rough runway combined to unleash the rocket side of this rocket car. The lightweight front end began bobbing up and down. Everything happened too fast for Tatroe to correct the situation. The car veered from the runway, bounced off the ground, and flipped six times in midair before crashing in a cloud of steam and car parts. "Man, what a ride!" Tatroe remarked, after emerging shaken but unhurt. The car was later rebuilt and had a long career entertaining crowds at drag strips.

In 1966, when Bobby Tatroe survived that very brief rocket ride, Robert C. Truax had been in the rocket business for nearly thirty years. That didn't even count those youthful experiments with gunpowder rockets or that interval when he would find discarded movies in the trash behind the University of California. If you didn't burn down the neighborhood or blow yourself up in the process, cellulose nitrate film stock just happened to make a most excellent rocket propellant.

That was the early 1930s, and what with the activities of American rocket pioneer Robert Goddard and the newly formed American and German rocket societies, the occasional article would appear in such magazines as *Popular Mechanics* and fill young Truax's head with visions of rockets.

By the time Truax entered the U.S. Naval Academy at Annapolis in 1935, Robert Goddard had moved his rocket experiments to the New Mexico desert, and the American Rocket Society (ARS) was conducting its own experimental work on liquid fuel rockets. "The mere thought of mankind exploring the wonders of the universe fascinated me. I also liked mechanical gadgetry. So, I read everything I could find on the subject," Truax recalls about this period in his life. When he became an upperclassman, reading gave way to experiment. In his spare time, Midshipmen Truax hung out at the Steam Engineering Building, where he could use the metal lathes to fabricate parts for his own rocket experiments.

This was not the sort of spare-time activity that midshipmen generally engaged in. He had to wheedle the use of machinery, metal stock, and testing locations, squeezing out permission from doubtful superiors. With the help of a sympathetic welder, he built a test stand and a combustion chamber fueled by gasoline and compressed air. Start the spark plug, feed in the air, then the gasoline, adjust the valves. At first it pop-popped like an out-of-tune car engine. But after numerous trials, he found the right combination of fuel and pressure to smooth out the combustion and produce a roaring, controlled thrust.

Truax marked the beginning of his rocket career from that moment, on 20 December 1937, the day he made his first engineering measurement on a rocket. The *Journal of the American Rocket Society* published his description of the experiment in its April 1938 issue.

When Robert Truax came off two years of ship duty in 1941, he was assigned to the Engineering Experiment Station (EES) at Annapolis. He was the only person in the Navy who knew much about rockets, and what the Navy needed rockets for was to lift its PBY seaplanes out of the water. The large planes were underpowered and with any kind of payload struggled to get airborne.

Truax prepared a report on the various power-assist options available and who might do the work. The Navy could contract the work out to Robert Goddard, he reported, or to the group at Caltech that had been experimenting with rockets, or they could do the work in-house. A fourth option was to have the Virginia-based National Advisory Committee for Aeronautics (NACA) do the work. When one senior engineer quipped that "a program set up within the Navy and run by a peppy and enthusiastic young naval officer would accomplish more in a year than the calm deliberations of the NACA could do in a century," the course was set.

Truax was delighted. "I wanted to get out of the paper-shuffling environment and get my hands dirty again." He was put to work developing a JATO rocket to give the PBY a boost. The term "rocket" still carried the air of quackery; hence the use of the more conservative term, "jet propulsion." His first order of business was to reconnect with everything that had gone on in rocketry while he had been at sea.

He visited Caltech, where the scientists were conducting basic research in combustion and cooling at their new test site in the Pasadena hills. This group would give rise to the Jet Propulsion Laboratory (JPL). He attended the test firing of rocket engines by the ARS. The ARS had been formed in 1930 as the American Interplanetary Society (AIS) by G. Edward Pendray and several other science fiction writers, but in the early 1930s AIS membership swelled with scientists and engineers less intent on going to Mars than on solving the mechanical problems of rockets. In 1934 AIS renamed itself the American Rocket Society and began its own serious rocket testing program. Truax had joined in 1936 as member no. 306.

Meanwhile, back in Truax's EES office, who should show up one day but the legendary Robert Goddard. He had been working under a grant from the Guggenheim Foundation, but with war approaching had been urged to work for the military. If he was to make more progress on his rocket experiments, he told Ensign Truax, he would need the help of a good engineer. "I asked him if he had anyone in particular in mind," Truax recalled in his unpublished autobiography, "and he said that he was thinking of me."

Although Truax greatly admired Goddard, he turned him down, in large part because of Goddard's personality and the secretive approach he took toward his research. Truax and the men on his crew were enthusiastic about their work and loved to talk about it and share ideas. Goddard did not. Truax felt that he would not be a participant in the research with Goddard but merely an assistant helping to develop Goddard's ideas. Truax kept his independence, but at the same time a separate, parallel JATO development program began under Robert Goddard.

It wasn't long before a dictum came down that whenever Goddard was about to run a test, everyone had to adjourn to the upper floor of the nearby Metallurgy Building and pull down the shades so as not to observe it. Truax knew he had made the right decision. Such practice was as opposite to his approach as one could get. A precocious engineer gushing with ideas that he burned to share with anyone who would listen, Truax made rapid progress building his own JATO program.

The U.S. entry into World War II gave the project greater urgency and brought in new staff. Twenty-five engineers would eventually work on a team of 120 civilians and 100 enlisted men. Within eighteen months they had designed, developed, and flight-tested recoverable JATO units, which were then mass-produced as part of the war effort.

It was a remarkable accomplishment, an experience that crystallized in Robert Truax many of the abiding principles that governed the rest of his career. One clear lesson to be taken from the experience: Driven individuals were better at pushing the development of rocketry than was the bureaucracy that controlled them.

It's hard to say how profoundly the launch of a single rocket on 16 April 1946 affected the American rocket community. It certainly made an impression on Robert Truax, who was at White Sands Proving Grounds, New Mexico, for the occasion. He had already interviewed Wernher von Braun, one of the German rocket scientists now working for the U.S. Army, about the production of v-2 rockets during the war. Now he was going to witness the launch of the huge rocket.

The Army had constructed a massive blockhouse that couldn't accommodate the overflow crowd gathered that day for the launch of the first v-2 rocket in the United States. Truax found a vantage point behind a corner of the blockhouse where he had a close-up view of the massive rocket. A little too close, actually: immediately after launch the rocket leaned toward the blockhouse and began corkscrewing upward. "I was frozen in my tracks. I watched with utter fascination as the fifty-foot rocket ascended slowly, at one time right over my head so that I could see the big jet of flame end-on."

The whole experience was enough to leave even a level-headed engineer like Robert Truax with the impression that he was squarely in the middle of a transformation in the thinking about rockets. That giant rocket twisting directly over his head was blasting open the door to the space age that might soon see men voyaging among the stars. And he wanted to be part of it.

The two decades of Robert Truax's active career, 1939–59, certainly spanned the opening years of the space age, from before the development of the v-2 to the creation of NASA and the launch of the first Soviet and U.S. satellites. And yet, as Truax came to realize, the door to that new age had not exactly been blasted open on that momentous day in New Mexico; instead, it had only begun to swing open, at considerably less than rocket speed.

As was typical for the hard-charging Truax, his impatience on the issue of space ran ahead of prevailing attitudes, making him appear to be either a visionary or a crackpot. Though the v-2 had brought respectability to the rocket, rocket development was almost exclusively focused on the military rather than on space. In 1951, when the ARS awarded Truax its Robert H. Goddard Memorial Lecture Award for outstanding achievement in the field of liquid propellant rockets, he began a long struggle to transform the society into an advocate for space.

Facing an audience of rocket luminaries at the ARS dinner for his award, he used the occasion to berate them. As an organization that should have the greatest influence in advocating for the U.S. entry into space, the society was doing precious little. "Stop talking about space travel and start doing something about it," he admonished. He proposed the formation of a space flight committee to explore the technical and political feasibility of creating a program to explore space. Although he received polite applause, talk of space was still seen as the domain of crackpots and fiction writers. In the audience that night, Truax's wife Rosalind shared a table with Richard Porter, head of the Army's Hermes missile program, who remarked, "Well, I, for one, will keep my feet on the ground."

Military rockets had always trumped space rockets. Even thinking about space could get you in trouble. Von Braun had been arrested by the Gestapo in 1944 for allegedly being interested in using the A-4/V-2 for space travel. Ideas had not changed that much in the intervening years.

In that postwar decade of interservice rivalry over rocket development, the Navy was out of the picture. The Army had Von Braun's ex-Peenemünde group laboring on a medium-range ballistic missile at the Redstone Arsenal in Alabama, it had the Hermes program with General Electric, and it had already developed the Corporal rocket at JPL.

The Air Force was intent on building the biggest rockets, intercontinental rockets to replace its long-range bombers in the delivery of nuclear warheads. To fast-track development, the Air Force had created a separate research and development unit, the Western Development Division (WDD), commanded by Brigadier General Bernard Schriever.

Meanwhile, the Navy had Truax working on the Regulus, a small jet airplane/cruise missile fired from the deck of a submarine.

Truax had proposed making the ballistic missile the primary weapon aboard submarines, but his proposal didn't get anywhere. It was more than any self-respecting rocket engineer could take. Truax felt a building frustration that he was very much out of the action.

One day he took a stroll in the "E" ring of the Pentagon to offer his services to an old acquaintance, Trevor Gardner, then in charge of Air Force research and development. "Ever since we started this program, I have been wondering when you would be coming through that door," Gardner remarked. As fast as it could be arranged, Truax was put on official loan to the Air Force, where he would stay for the next three years, 1955–58, lead-ing the design effort on the Thor missile and later serving as deputy direc-tor of the super-secret reconnaissance satellite program, which would even-tually produce the Discoverer, Midas, and Samos satellites.

Truax, who usually chafed at administrative oversight, was reasonably happy at WDD. As he would so often characterize ideal working conditions, "They told me what to do but not how to do it."

When Truax became president of the ARS in January 1957, he seized the opportunity to advance his belief that the United States should be aggres-sively developing a program to explore and utilize space. He immediately appointed Krafft Ehricke to head a commission to make recommendations at the group's September meeting. His report called for the creation of a new space agency and made glowing predictions about how it would usher in an era of discovery and international cooperation. The ARS board of di-rectors, which included von Braun and top people from the Army, JPL, and defense contractor rocket programs, approved the report to be passed along to President Eisenhower. It arrived on the president's desk in mid-October, two weeks after the launch of *Sputnik*.

A month later, after the Soviets launched their second satellite, this time with a dog passenger, Truax begged Schriever unsuccessfully to let him have one of the backup Atlas rockets, with which he would put a ton of scientific hardware into space. It wasn't long after that that Truax returned to duty with the Navy to work on the Polaris program. But by mid-1958 he had found a spot more to his liking at the Advanced Research Projects Agency (ARPA), a Department of Defense space agency created in the wake of *Sputnik*.

When Congress created NASA in October 1958, assorted ARPA-NASA com-mittees came into existence to ease the transition to civilian control. It fell to Truax to serve on the Man in Space committee.

Although he had long championed the need for a civilian space agency, there immediately arose in him an uneasiness with the way NASA did busi-ness. In this instance, it was their disregard for cost efficiencies.

Max Faget served as NASA representative on the committee, and a disagree-ment quickly arose over which rocket to use for the Man in Space program. Faget favored the Atlas, while Truax supported an Atlas-Agena combina-tion, which was just finishing development. The Atlas-Agena was a larger rocket that would eventually be used to launch the Midas and Samos reconnaissance satellites. To Truax's way of thinking, it would have served both for launching the Mercury capsule and for launching the larger, two-man Gemini capsule to follow.

But Faget wanted the simplest route to a Project Mercury launch vehicle, and his argument won out. The Atlas was developed for Project Mercury and the Titan II for Gemini. "We could have done it all at one time with an Atlas-Agena, and saved a couple hundred million dollars," according to Truax. But no one was counting the millions in 1958.

As Truax approached the end of his twenty years of service in the Navy, he concluded that there were no positions available that appealed to him. He put out his résumé to private industry with the simple stipulation that he have a million-dollar budget to use as he wished. Aerojet accepted the offer. His move from Washington DC to Aerojet headquarters in Sacramento, California, also happened to coincide with a divorce from his first wife, Rosalind, and marriage to his second wife, Sally.

Things were buzzing at Aerojet-General's Liquid Rocket Plant when Truax showed up for work in 1959. They were producing rocket engines for the Titan missile but had been remiss in not spending that small percentage of their contract allowable for independent research and development. That would be Robert Truax's job. He was given his million dollars and put in charge of the Advance Development Division.

With a free rein over the course of his work, he struck off in two directions: determining cost factors for space transportation and building a pressure-fed rocket.

"For years I had been bugged by the high cost of space transportation," was how he explained his motivation in his unpublished autobiography. "Nobody really knew how much it cost to put a pound of payload into orbit, or what factors dominated the cost picture."

In 1954 he had presented a paper at the International Astronautical Federation conference in Zurich titled "Economics of Satellite Supply" that drew on the technical detail in von Braun's 1949 novel, *Project Mars*. A winged spaceplane crowned the top of von Braun's fictional, three-stage Mars rocket. A rocket with wings? It didn't make sense to Truax. A tail, wings, landing gear: that was all dead weight for a spaceship, weight that displaced payload. It wasn't worth it just to recover the rocket's final stage.

He put a team to work gathering and analyzing available cost data on rockets, and they came up with some basic recommendations: (1) make launch vehicles reusable, (2) make them big, (3) make them simple, and (4) don't mix people and cargo.

If much of the cost of launch was in the hardware and it got destroyed whenever you used it, then reusability made sense. The study also showed that the cost of a rocket varied less with size than with complexity. The more parts it had, the more expensive the rocket. There was no need to push the state of technology to the limit. More sophistication meant more parts, which translated to higher cost. And if you really wanted to jack up the cost, build it to the level of reliability required for human passengers. Cargo did not need that level of reliability. Mixing humans and cargo meant that you had to build that very expensive reliability into the entire vehicle for just the miniscule part of the cargo that was human payload.

Before Truax's team entered the design phase for their pressure-fed rocket, they checked in with the Future Projects Branch at Marshall Space Flight Center in Huntsville, Alabama. Think in terms of a manned mission to Mars, they were told, something that could deliver a million pounds of payload to orbit.

For some perspective, the giant Saturn V rocket that boosted the Apollo moon flights had a payload capacity of 260,000 pounds. The Saturn rocket, still in development at this time, would reach its apotheosis with the Saturn V, a rocket of colossal dimensions: 363 feet high and 33 feet in diameter. A rocket capable of lifting a payload four times greater than the Saturn V would have to be of truly gargantuan proportions.

The rocket that Truax's Aerojet team eventually designed broke the mold in numerous ways. It was a quantum leap in size above the Saturn. Named Sea Dragon, it stood more than five hundred feet high and was seventy-five feet in diameter, with a launch weight of four million pounds and a payload of over a million pounds. But to dramatically reduce the costs of transporting payload to space, Truax also turned his back on the delicate, ultra-sophisticated NASA designs in favor of industrial-grade construction and performance.

Sea Dragon would be a rugged, two-stage rocket that used a single, pressure-fed engine per stage. It was literally too large to be built and transported on land and would have to be constructed in a shipyard. This was no problem, because Truax was redefining rocket design on this project. Sea Dragon's hearty tankage could be built as one would a submarine hull, using 8mm maraging steel to provide the strength for the high psi of the pressure-fed engine and the robustness to hold up to recovery and reuse. Built in water, Sea Dragon would also be launched from water. The rocket would be towed out to sea, where a water-filled ballast tank would stand it erect for launch.

Based on the concept of simple parts, industrial-grade production, a lower-performance propulsion system, and the reusability of key components, Sea Dragon was the original design for what would come to be called the "big, dumb booster." It was a space truck, a space "ship," designed to provide economical access to space. Aerojet's 1963 report to NASA pegged the cost of lifting payload to orbit with the Sea Dragon at $20 to $30 per pound.

When the design was finished, Truax's team made a presentation to the Future Projects Branch at Marshall. "Our presentation left them aghast and incredulous," Truax recalled. "The very idea that something less than the most elegant could actually be better somehow went against the grain." To NASA's surprise, all of Sea Dragon's design features and economies were confirmed through independent studies. However, the end of the Apollo program and the subsequent budget cuts eliminated any thought of a Mars mission and the need for gigantic rockets. Sea Dragon never passed from the design phase.

Sea Dragon's demise roughly corresponded to the appearance at Aerojet of Art Arfons, and Truax's decision to create Truax Engineering. The restless Robert Truax moved on to career number three.

Later on, there would be bountiful years, when Truax Engineering struck gold with a big Navy contract that revived the Sea Dragon concept. But originally it was a part-time effort that Truax pursued in his off-duty hours, whenever he had a project. Meanwhile, he continued to work on a succession of jobs that took him back and forth from the East Coast to the West. There were brief stints with the Air Force, a joint Navy–Department of Commerce project on hover craft, and the Minuteman program. "I had become a paper-pushing bureaucrat with a persistent yen to get my hands dirty again."

Truax's official retirement came in December 1973, although it certainly mischaracterizes his situation to use the word "retirement." He was "not currently funded," is how longtime Truax Engineering employee Dan Slater described the periods between projects, even as Truax continued to work into his seventies and eighties.

The Truax family settled into a home in Saratoga, California. The new house came with an appealing three-car garage that immediately filled with equipment and Truax's collection of surplus missile parts, a collection that spilled into the backyard, or "boneyard," as it was known.

There had been a time during the 1960s when someone could poke through the scrap yards of southern California and find the cast-off remains of America's ballistic missile programs. The Titan II had replaced the Titan I. The Atlas and Thor missiles had been decommissioned and their innards sold for scrap. Truax had followed up work on Walt Arfons's steam rocket car by building him a wheel-driven dragster that got an extra kick from the turbo-pump assembly out of a Titan I missile.

One day while treasure hunting, Truax came upon seven Atlas vernier rocket engines. The engine looked like a nickel-plated fog horn, but it was a sweet piece of craftsmanship: regenerative cooling, thrust chamber gimbaling, designed to run on kerosene and liquid oxygen. Manufactured by Rocketdyne for both the Atlas and the Thor, it developed one thousand pounds of thrust for attitude control on the missile.

Truax bought them for $25 apiece, a bit of technology that had cost the government about $70,000 each. You never knew when you'd need a rocket engine like that. They were added to the titanium pressure spheres and the turbo pumps and what-have-you, the vast collection that traveled like an albatross with the family from coast to coast as he moved between jobs.

When they settled in at Saratoga and the collection began to decorate the yard of their new house, Sally Truax put her foot down. In a colossal case of misjudging her adversary, she gave her husband an ultimatum: "Either the rocket parts go or I do." Scott Truax, Bob's oldest son from his second marriage, who had a clear grasp of his father's mindset, already knew that rockets were number one and family was a distant second. "Dad's reaction was 'Okay, don't let the door hit you when you leave.'"

Sally made good on her threat and left, but returned in a few weeks, reconciled to the priorities that ordered her husband's life. None of them was aware just how quickly the work of Truax Engineering would consume their lives.

An engineer named Doug Malewicki stood above a chalk drawing of a motorcycle he had just sketched on the cement floor at the American Motorcycle Company. Bob Truax studied it for a while. It didn't look like any motorcycle he had ever seen. Malewicki worked for a guy named Evel Knievel, who wanted to jump the Grand Canyon on a motorcycle. Sounded crazy, but Malewicki was thinking that some kind of fully enclosed cycle with a rocket assist might do the trick.

After some hasty calculations, Truax realized that it was just too heavy. "We've got to get rid of that reciprocating engine," he said. One thought led to another, and they soon concluded that what was really required to launch Knievel over the canyon was not a motorcycle but a rocket. Malewicki didn't think Knievel would go for it; he was a motorcycle stuntman, not an astronaut. But when he phoned him and got the go-ahead, Truax roughed out the basic design for the rocket on the spot.

Working with Evel Knievel in 1973–74 plunged Truax into an alien world. When Knievel and Malewicki had a falling out, Truax got sole control of developing the vehicle that Knievel would ride across the canyon. It was a mixed blessing. Truax liked to be in charge of a challenging project, but it also put him in the position of dealing directly with the manic, adrenaline-pumped world of the flamboyant stuntman.

Knievel was riding a tidal wave of public attention. American Eagle Motorcycle had just signed him to an endorsement deal. A movie titled *The Evel Knievel Story*, starring George Hamilton, was in production. Meanwhile, the media fed the public a steady diet of Knievel mania, featuring his motorcycle jumps on ABC's *Wide World of Sports* and closely following his high-profile legal case against the federal government to win permission to jump the Grand Canyon. Mix this world with the cerebral group of engineers and technicians working with Truax and you got a clash of cultures.

When the government nixed the Grand Canyon jump, the venue moved to Idaho's Snake River Canyon instead. Truax started from square one to analyze the problem. For starters, that twenty-degree mound of dirt Knievel had already piled up on the canyon rim for a motorcycle launch ramp would not do the trick.

When Knievel saw the steel launch rail that Truax constructed over the dirt, even his hardened, stuntman heart rose to his throat. "My God, that rail is going straight up!" he exclaimed. The rail stood at a fifty-one-degree angle. For the first time, it dawned on the daredevil just how different this stunt would be from all previous ones. He would not be driving a motorcycle but riding a rocket.

The rocket in question had acquired the name Skycycle X2 and it would be powered by the reliable steam rocket Truax had developed for the Arfons dragster. Operating it at a higher pressure would give it the additional thrust required. An old airplane wingtip fuel tank served for the airframe, the fins came from a helicopter, and an autopilot came from a surplus Nike missile.

Funding is always a critical issue for rocket builders, but Truax's experience with funding politics in the military and the federal government did not prepare him for being a rocket entrepreneur and working with a showman. He had a pay-as-you go understanding with Knievel, who proved most enterprising in securing funding.

"Bob, will that thing run on beer?" came a Knievel call one night. He had to be kidding, Truax thought, but Knievel kept pushing for an answer. "I'm here in Milwaukee; I'm with the Pabst people." So Truax conceded that if you heated the beer and boiled off all of the alcohol, it would probably work. But nothing came of it. Somewhat later, a call woke Truax one night at 2:00 a.m. Knievel was in the hospital in Seattle following a jumping accident. "I'm trying to cut a deal with Olympia Brewing. Is there any advantage to using soft water?" Olympia finally did sign on to put their name on the rocket's water supply.

Eventually, Knievel handed promotional duties over to Bob Arum, fight promoter for boxer Mohammad Ali. Arum sold closed-circuit TV rights, and suddenly preparation shifted into high gear. It was May 1974, and the jump had to take place in four months.

The building of Knievel's Skycycle set the tone for how Truax Engineering would operate on future projects. Truax put out the call to friends and family. A machinist that Truax had worked with on the East Coast took a leave from his job to join the effort. "We put him up at our house, and the poor devil worked twelve hour days, just for the love of it." Several engineers from Lockheed joined the team. According to Truax, "What followed were wondrous days. We never really knew how long the project would take or what kind of difficulties we would encounter, so no effort less than the maximum would do. We worked until we were ready to drop. We all felt that we were part of a great adventure."

The team built two copies of the Skycycle x2. A static firing and two test launches over the canyon revealed problems that had to be fixed. Truax pleaded for a delay, but Knievel wanted to hold to the September 8 launch date.

A carnival atmosphere prevailed at the site on launch day. ABC Sports was on hand, as were BBC TV journalist David Frost and astronaut James Lovell, as well as some thirty-five thousand spectators. After considerable ceremony, Knievel got on board. Five thousand pounds of steam thrust shot him into a high arc above the canyon. A gasp went up from the crowd of spectators on the canyon's south rim.

An unplanned deployment of the parachute interrupted the Skycycle's clean trajectory. Even with the chute fighting the thrust, the rocket made it across Snake River Canyon, but wind caught the chute and drifted the rocket back into the canyon. The Skycycle banged against the canyon edge before the chute lowered Knievel to the floor. He had survived with only minor injuries.

Truax would eventually take the blame for the chute malfunction, but the incident got overlooked in the postlaunch thrill of the moment, when Knievel remarked, "Bob, that is going to be one hell of a hard act to follow. What else have you got up your sleeve?"

Truax had already been thinking in this direction. Working with Knievel had impressed upon him how much money could be raised to finance a crazy stunt like jumping a canyon — promotional money, TV rights, corporate sponsors. So when Knievel posed his question, Truax answered with his idea for an even crazier stunt. "If you can raise a million dollars, I think I can make you the world's first private astronaut."

In September 1974, the same month that Evel Knievel rode a rocket into the Snake River Canyon, Gerard O'Neill's article "The Colonization of Space" appeared in *Physics Today*. O'Neill's colonies were predicated on cheap access to space, having something available like the shuttle to serve as a space truck, carrying bulk cargo to orbit.

It didn't work out that way, of course. In fact, it was the disappointing cost of shuttle cargo delivery that dealt the final blow to O'Neill's bold plans. Originally projected at $3 million per flight, the shuttle ended up costing more than $1 billion per flight. Each flight of the reusable shuttle turned out to be more expensive than the expendable Saturn V.

From the beginning, the whole idea of the shuttle stuck in Truax's throat. "Wings on a rocket made about as much sense as tits on a bull," was his oft-repeated quip. Throughout the 1970s and '80s, Truax seemed on a mission against what he viewed as NASA extravagance and its misguided devotion to a winged spacecraft. In a 1970 article, Truax called the shuttle an "unparalleled money sponge." He argued that using existing technology, such as Gemini or Apollo ballistic craft, would serve just as well and would dramatically speed up space development and reduce costs.

The point was that he could do it better, for less money, and now that he had seen Evel Knievel's ability to raise money, he too could find funding to launch his own private rocket. The obsession consumed him for a decade.

Truax's quest to build a private rocket began with the promise of backing from Knievel, but a brush with the law robbed the daredevil of his fundraising ability. As Truax saw it, this left him with two problems: (1) where to find enough money to continue the work and (2) where to find "another person crazy enough to ride a homemade rocket."

When an article appeared in the San Jose, California, newspaper about Truax's plans to launch a private astronaut, he quickly realized that no. 2 was not going to be a problem. More than four thousand letters arrived from eager volunteer astronauts, including engineers, airline pilots, a grandmother, and a Buddy Holly impersonator. Plenty of people also offered to help build the rocket. But no one appeared with the few million dollars needed to fund the project. In the years to come, he would work his way through a bevy of fund raisers, PR men, would-be investors, scam artists, sincere businessmen, and astronaut candidates with private funds.

"Wanted Venture Capital for Risky Venture," read the advertisement that Robert Truax placed in the *Wall Street Journal*. That bit of honesty in advertising got no response.

But if the Wall Street money people lacked interest in the project, not so a host of less reputable sorts who dabbled in speculative investments or who promised to work on Truax's behalf to find the money people. In fact, from the mid-1970s to the mid-1980s, trying to raise money consumed a good deal of his time.

People literally showed up at his door offering sweet financing deals that never materialized. On a few occasions he met in hotel rooms with shady characters who talked about large sums of secret, off-shore money that needed a short-term investment. Would-be astronauts appeared with small infusions of cash.

An ex-tax adviser from Los Angeles, working on Truax's behalf, raised $100,000 from a local businessman. The guy rented an office and hired a secretary, then rented the ballroom at the San Jose Hyatt hotel to throw a big "coming out" party, complete with TV monitors to show film of rocket test firings and slick brochures outlining the various investment options. Ten people showed up, Truax's friends mostly, plus one or two potential investors. No one invested.

Truax called his rocket the Volksrocket, a name derived from the inexpensive, German-made car popular at the time, the Volkswagen. "Volkswagen" means "people's car." The Volksrocket would be the people's rocket, offering the same economical access to space that the car offered to the roadway. Truax built a dummy version of the rocket for the adviser to display at Los Angeles shopping centers to raise money. He didn't.

That failed effort got logged into the loose-leaf notebook in which Truax kept track of his fundraising attempts. One bulging notebook grew to two, then three. "Dad was never good at chasing down money," claims Scott Truax. "He was all about rockets. He had the idea that if you built a better mousetrap the world would beat a path to your door, which is absolutely false. It will beat a path to the door of the guy who has the best way to market his mousetrap."

But Truax had studied freelance fund-raising at the feet of Evel Knievel and had developed an appreciation for the power of the media to draw attention to a cause. Throughout the 1970s local and national media continued to cover the story of the rogue rocket engineer who planned to launch his own private astronaut.

Overall, the story went something like this: Legendary rocket engineer plans to build a rocket with private financing to carry an astronaut to an altitude of fifty miles. For safety's sake and to keep down costs, the rocket uses parts from other successful programs, including vernier engines from the Atlas, gyros from the Nike Hercules, as well as parts from the x-15 and Polaris. In the cramped pressurized nose cone, the astronaut will have enough air for thirty minutes. The rocket will expose the astronaut to 3 g as it lifts him to the edge of space, where he will experience weightlessness and be able to peer through a small window to see the blackness of interstellar space and the curvature of the Earth below. TV will capture the rocket flight, and through a communications link, the astronaut will describe the experience to viewers. Recovery will be by a parachute water landing after a total elapsed flight time of ten minutes.

This played well with the public in the post-Apollo 1970s, when NASA's great push into space had stalled and all attention had shifted to development of the space shuttle. Here was a guy with genuine rocket credentials—a visionary, some said—who was saying that NASA had it wrong and that the shuttle was an overpriced, bad idea from the get-go. Private industry ought to take the lead in space, Truax thought. And he intended to prove that it could be done, right now, using 1960s technology and a modest amount of private money. He was going to send a man into space.

The media put a Don Quixote spin to the story, calling it "Bob's space program," run out of a garage workshop with secondhand parts. Truax loved it. He had managed to recapture a small part of the media attention that had surrounded the Knievel event. Surely financing would follow.

An article about Truax appearing in the October 1979 issue of *Omni* magazine caught the attention of Chicago real estate developer John Oelerich. Oelerich, a space enthusiast and former member of the L5 Society, already nurtured a frustration with NASA's handling of the space program. He called Truax, then visited his home. "Bob was a playful, funny guy, but I was impressed with his intelligence and that he was planning to do this with secondhand parts."

Oelerich saw the potential in a business partnership with Truax. Oelerich would give Truax what he had lacked for the past three years: financing and a businesslike approach to realizing his dream. Truax would give to Oelerich a chance to make a profit and more importantly a chance to make history.

"Over the past three weeks I have been working on one of the most unusual, exciting and potentially rewarding deals any of us have ever come across," Oelerich wrote a letter to potential investors in December 1979. "The situation involves one Robert C. Truax, a most fascinating and unique individual. For the past 3½ years, Mr. Truax has been proceeding, essentially on his own, with a project never before attempted, the development of a one man, retrievable rocket to space."

Investor response brought in an initial $240,000 in funding. Oelerich's plan made perfect business sense: provide Truax with enough money to continue design and testing of his rocket, while Oelerich handled the numerous business challenges.

Initial investor funding would not be sufficient to see the project through to completion. The estimated cost of rocket development was put at about $1 million. Yet it seemed reasonable to expect that amount of money was within reach. After all, they were creating a product of historic importance — the first private manned space launch. If ABC had paid $3 million for the rights to Evel Knievel's canyon jump, imagine the revenue potential for launching a man into space. Network TV, cable, magazines, book deals, foreign media, toys, product sponsorships: the marketing potential was unlimited.

For everyone involved, 1980 would be a very fast, adrenaline-filled year. The project was incorporated as Project Private Enterprise (PPE) because it would demonstrate that private industry had the means to develop its own manned space program. Oelerich moved quickly to pull together a management team and devise a business plan. PPE letterhead carried the slogan "lift off . . . fall 81," committing to a challenging eighteen-month timeline. Two key pieces had to fall in place to make it work: Truax had to maintain steady progress on rocket development, and Oelerich had to bring in additional money before investor funds were depleted.

Oelerich's monthly update to investors began in December 1979 and reported rapid progress. On 7 January 1980 he was able to report the hiring of John F. Feeney, NBC's former director of promotion, to handle marketing and public relations. The strategy was to first generate a lot of media exposure for Truax to give the project credibility and build interest. With that accomplished, they could put together broadcast, sponsorship, and merchandising deals and use that advance money to finish development of the rocket.

Feeney would be able to market three different stories: Truax himself, the rocket, and the astronaut. Because he was a legendary space engineer and pioneer, Truax's background lent the project its all-important credibility. In order to get advance commitment, people had to believe that the launch was actually going to take place.

The rocket, now given the name x-3, provided the visual impact for media stories. PPE created several videos about the project to use in publicity and negotiations when selling rights or seeking sponsorships. The stress here would be on the development efforts, problems, and accomplishments, all following a scripted timeline.

Although the management team gave considerable thought to astronaut credentials, the top qualification was the astronaut's promotability. A confidential memorandum told investors, "It is the management's opinion at this time that the astronaut should be a 'common, ordinary' person, one without pre-existing fame." Although they did not feature an astronaut candidate in their promotions, the person to fly in the rocket had already been selected, an engineer named Jeana Yeager, who already worked for Truax Engineering. Aside from her technical background and her willingness to do it, she happened to weigh only ninety-five pounds.

The media eagerly took up the PPE story. Articles were soon pending in *Esquire* and *Newsweek*, and also in British, French and German magazines. Taping dates had been set for *The Dinah Shore Show*, the *Tonight Show*, and ABC's *That's Incredible*.

If the media exposure was unfolding as planned, rocket development did not exactly follow the script. Oelerich had a hard time getting Truax to commit to a meaningful rocket development schedule. Those expected to provide advance money for broadcast and other rights needed a firm launch date. Truax's credentials were one thing, but could he actually deliver on this backyard rocket venture?

Eventually, PPE laid out a timeline that included several private static tests, then, when things were working well enough, a public static test for the media to generate big publicity. Development of all systems would proceed from there to an unmanned launch of the rocket in June 1981, to be followed by the actual launch in the fall.

Meanwhile, Oelerich and his team scrambled to bring all of the other pieces into place: finding a launch site, securing affordable liability insurance, nailing down a toy deal, locating a specialty steel for the rocket's pressurized fuel tanks. Although the project was under budget, investors were alerted that the second installment of their investment would be due in August. Also, they should mark their calendars for the big Media Day public static test of the rocket.

"Five, four, three, two, one." The crowd at the airport in Fremont, California, chanted the countdown along with Robert Truax during the static firing on Media Day, 24 June 1980. Then a ten-foot cone of flame blasted out the rear of the rocket with an ear-splitting roar, ricocheting glowing debris onto a grassy hillside one hundred feet away and starting small fires. Sixty seconds later the engine sputtered to a conclusion, and the crowd cheered.

Looking at the expended x-3 on its test stand, it was hard to imagine it lofting an astronaut into space. It was a modest twenty-five feet in length and twenty-five inches in diameter. The operation that supported the rocket was equally modest. A moving van served as mission control, Truax's Chevy El Camino had transported the rocket to the site, and the ground crew sheltered behind a line of fifty-gallon steel drums topped with sandbags. And yet, as PPE had planned, the event shifted media attention into overdrive.

"I'm very pleased with the way things went," Truax told the more than 150 reporters and photographers on hand. The national TV networks had cameras rolling. Reporters from many of the major city newspapers jotted down his words and snapped photos of this sixty-three-year-old rocket engineer who looked like a holdover from the *Sputnik* era, with his military-style flat-top haircut and olive drab outfit.

About ten o'clock that night Feeney got an urgent call from his telephone call center. He had worked a deal with an LA radio station that asked for listeners to call in and volunteer to be PPE's astronaut. "Feeney had hired three women to take the calls," Oelerich recalled. "But at the end of the day, they called him and said they were shutting down and going home. Their phones had not stopped ringing the whole day, and they had gotten 900 volunteer astronauts."

"Suddenly the continuing saga of Robert Truax, Space-Age Don Quixote, had taken on a certain respectability," commented the *San Francisco Chronicle*. Bill Geist of the *Chicago Tribune* wrote about the confidence the PPE team had in their ability to pull off the launch. Geist admitted that he too believed it would happen. But he hastened to note that he believed because he wanted to believe. "For truth, justice, and the American Way. For every weekend wrench jockey. For every taxpayer who winces at his money being flushed away. . . . For every one of us who has ever wanted to bust into a meeting of our supervisors and tell them to stop yakking and — Dooo! — something for a change."

Three days later Truax appeared on the *Johnny Carson Show*. In hindsight, Oelerich thinks that Media Day was a big mistake. Suddenly national TV shows were calling, and you couldn't turn them down. Truax began spending more time promoting the project than developing the rocket. It was a serious miscalculation.

"I spent most of August 1st with Bob and emerged with the realization that the rocket development schedule is in more trouble than I had sensed." Oelerich's gloomy words to investors in a 7 August letter reported that it was highly unlikely that PPE would be able to meet its June 1981 unmanned launch date. This was bad news indeed. Falling behind schedule meant more money, lots more money. More critically, it meant loss of credibility, which meant less chance of raising more money.

"Things got ugly very fast," Oelerich recalls, referring to the financial crunch that faced investors that August when they would have to ante up the second half of their financial commitment. The writing was on the wall. Rocket development was behind schedule, with no definite launch date. Oelerich pressed Truax to pick up the pace, but he couldn't or wouldn't be hurried. Nor was financial relief in sight that might have allowed for a more leisurely schedule. Despite phenomenal media exposure, no advance money appeared from broadcast rights, publishing rights, or merchandizing deals. Oelerich and his Chicago investors cut their losses and got out of the rocket business.

For the next few years, Truax's preoccupation once again became the search for money. The pages in his loose-leaf notebooks of failed funding appeals grew at an alarming rate. "I wrote to every multimillionaire in the country that had a history of backing weird deals." All the foundations that sponsored such research got letters from him, as did a handful of major corporations.

His appeals seemed to play better overseas. A plywood company from Indonesia and a Japanese importer showed interest. They corresponded, they visited, but nothing came of it. He sparked the interest of Saudi Prince Turki al-Faisal. Truax visited Saudi Arabia and together they wrote a couple of proposals for funding to the Saudi government, without success. Finally, in media interviews, Truax began saying that anyone who raised at least $100,000 for the project would move to the top of the astronaut candidate list.

It seemed like a golden opportunity to a guy named Fell Peters. Born in 1958, he had grown up in the era when astronauts were cultural heroes, when larger-than-life figures orbited the Earth and walked on the moon, when every red-blooded kid wanted to be one. Although only twenty-five years old, Peters was already an enterprising businessman who was trying to make his fortune just then selling cat rocks, rocks with pictures of cats on them that capitalized on the earlier popularity of pet rocks.

Peters didn't have the money, but he had chutzpah. He laid his hands on $40,000 and converted it all to hundred-dollar bills. He bundled them neatly into thousand-dollar stacks, then he visited the rocket guy. Sitting at Truax's kitchen table, Peters laid out his qualifications to be an astronaut: young, healthy, single, and a good athlete.

He didn't have the full $100,000, he confessed, but he had a backer who would come across with the rest. When Truax asked exactly how much money he did have, Peters began to dramatically count out the stacks of hundreds. It was a nice touch. Truax had already decided not to accept anything less than the full amount, but as the pile of cash started climbing, he wavered. By the time Peters reached $40,000, the job was his.

For the cash, a promise to raise more, and a commitment to work at least half-time for the company, Peters received a two-year option to be the astronaut and a 6 percent stake in the company. The year was 1982, and PPE was at a low point. Things had fallen through with the Chicago investors, funding was nonexistent, development had stalled, and Truax was depressed. If they could get things rolling again, if they could finally sell broadcast and other rights, Peters figured he'd make a few million bucks.

Plus, real money also awaited the company that could launch commercial satellites. After PPE made history with its astronaut launch, it would turn its attention to creating a rocket named Excalibur for low-cost satellite launches. A scaled-down version of the massive Sea Dragon that Truax had developed at Aerojet, Excalibur would undercut the competition by following the same big dumb booster principles of simplicity and industrial-grade production.

In many respect, Peters's plans mirrored those of Oelerich: give Truax funding, professionalize his backyard operations, and find additional money to carry the project to completion. Peters began by cleaning up and organizing Bob's cluttered garage, hanging a sign out front, creating a new logo and brochure. His official duties were twofold: (1) handle public relations by making contact with potential investors and sponsors and writing press releases and (2) train to be an astronaut. His enthusiasm brought energy and focus back to the project.

To prepare his astronaut for the historic flight, Truax took Peters up in his VariEze, a Burt Rutan experimental airplane that he owned. Peters seemed to relish the 4-g pull-ups and turns Truax threw at him. He would later earn his pilot's license. Truax also enrolled him in a parachuting course, and he excelled in that as well, eventually recording more than one hundred jumps. Someone from NASA Ames Research Center who knew about the project invited Peters to use some of their astronaut training equipment. He did some negative-g training there.

Media interest in PPE was still intense. PPE was still enough of a novelty that the public wanted to follow the story. "We had a dog-and-pony show for the TV crews," recalls Craig Adams, a PPE engineer. "We'd have a rocket or two on display, on a transporter or test stand. We'd show them how the engines gimbaled. We'd show a video tape of static tests. Fell Peters would demonstrate how he'd be crammed into the rocket.

"They always had the usual questions: 'Why are you doing this? How can you do it so cheap? What is NASA doing wrong?'" This gave Truax the opportunity to explain that you didn't need all NASA's fancy equipment to launch rockets. And you certainly didn't need wings and landing gear on your space vehicle.

Truax continued to be a magnet for those inspired by his vision. If a volunteer hung around long enough, Truax put him to work. Craig Adams was in junior high when he first contacted Truax for advice on building a small rocket. After getting a degree in aerospace engineering and working for Ford Aerospace, he left that position to work for Truax.

Every new magazine article or bit of TV coverage brought in letters from well-wishers. "After one ABC show, we got over 300 letters. People sent money, even people from England and Japan," Peters recalls. An English teenager named Trevor kept sending letters with a few pounds of his hard-earned money in each.

People wanted to believe in PPE; they wanted to be part of it in whatever small way they could. A furniture store in San Francisco donated beds; a sporting goods store thought a trampoline would help in astronaut training. An up-and-coming Apple Computer was just three miles down the road. Apple founders Steve Jobs and Steve Wozniak visited PPE operations and were so impressed they sent a truckload of computers the next day.

A number of NASA's astronauts took a keen interest in the project. Alan Shepard, Wally Schirra, and Deke Slayton would write or call, wondering when PPE would launch. Peters wrote to British adventurer/entrepreneur Richard Branson, and invited him to invest in PPE. Write again when you have launched the rocket, Branson said.

Additional infusions of cash from Peters and from another investor kept rocket development chugging along, but there was never enough money. Bob still followed the government surplus lists and scavenged the private surplus stores for parts. "We never bought a new part when a used one would do," was how Truax defined his approach. "We never designed a part from scratch if we could find an 'off the shelf' part that would do well enough. We never paid for anything we could get for free. 'Good enough' was always good enough."

They did numerous static firings, built a ground control station, and a launcher. The astronaut capsule now sported complete controls and a parachute system. When they needed a transporter/erector, Truax contacted a local trailer company, which built it for free.

By the mid-1980s Peters was desperate to launch and pressed Truax to speed up development. Although they managed to work together well enough, various issues heightened tensions between the two men. Something as simple as the creation of a new brochure would involve a several-month battle of wills.

Truax's openness about the project also troubled Peters. Even when Truax used existing technology, he would sometimes add a new twist or improvement to it. "I'd tell him you need to get a patent on that, and he'd say we don't need a patent. He didn't want to spend the money. 'You've got to keep certain things secret,' I'd tell him," Peters recalled.

Case in point: the Rutans. Both Burt and Dick Rutan were good friends of Truax. Burt had founded a company named Scaled Composites in 1982 to develop experimental aircraft. Truax would visit them in Mojave, California, and they would come to his house for barbeques. They always came for static firings. Peters claimed, "Bob would tell them everything we were doing. Burt would be very quiet and take in as much as possible. He would ask key questions, and I'd think, oh no, he's going to compete with us. But when I'd tell Truax, he'd say, 'No, he's into airplanes; he's into the *Voyager*.'"

Ironically, one of the pilots for the *Voyager*'s historic 1986 nonstop, round-the-world flight was Truax Engineering employee Jeana Yeager.

Although Peters had been desperate to launch in 1985 or '86, the rocket was just not ready. According to longtime Truax Engineering employee Dan Slater, the propulsion system was in good condition, but "we were missing the recovery system, much of the vehicle avionics and associated ground support equipment, and more mathematical modeling was needed to insure vehicle stability and proper flight characteristics." Maybe another million and another year would have done it. Maybe not.

What was becoming crystal-clear to Truax, after a decade of hand-to-mouth existence, was that he was not going to be able to recreate the Evel Knievel path to project financing. Maybe Knievel could stroll into a corporate office and wheedle a sweet promotional contract, but Truax couldn't. Maybe the TV networks would put millions on the line for a Knievel jump because they knew he could deliver. They could put the event into the fall schedule on a date certain and be guaranteed that the showman would be there in dramatic fashion to launch his motorcycle over his latest obstacle.

During an appearance by Truax on the David Letterman show, Truax showed video of a static firing of the rocket. Always the comedian, Letterman joked that he thought the idea with rockets was that they were supposed to actually go up into the air, not just sit on the ground and blow flames. Truax thought Letterman was making fun of him and was offended, but it hit on a critical issue with PPE.

For most of the 1980s Truax had been getting extensive press coverage. Truax was the Don Quixote of rocketmen, complaining about NASA, saying private enterprise should have its own shot at a space program, that it could do a better job without wasting so much money. He was the misunderstood genius, champion of every little guy who tinkered in his basement and found a better way to do something.

But at least to the viewing public and to the corporate types and media moguls, Truax had not delivered much, certainly nothing that would motivate them to put up front money for the rights to something that didn't appear to be happening.

"The only reason I planned to put a person in the rocket was to pay the bills," Truax would later confess. "A manned flight, I reasoned, had enough money-making potential to interest investors. I suppose it was the genesis in the Knievel Canyon jump that set this approach as my primary one to raise money."

He needed a new strategy. Actually, Truax didn't care what was put into his rocket. He wanted to vindicate his big dumb booster theories and prove the practicality of water-launched rockets. What these approaches promised was low-cost space launch capability. So, in the proposals he sent to foundations and corporations, he played down the private astronaut angle and stressed that his ultimate goal was to create a reusable rocket for low-cost space launch capability.

After all, the age of the communications satellite had dawned, and such companies as Western Union, Comsat, RCA, and Satellite Business Systems had already launched numerous private communications satellites aboard NASA rockets. Yet the shortage of satellites remained acute. NASA's rockets were booked solid for satellite launches, as were the flights of the shuttle. NASA charged in the neighborhood of $28 million to place a satellite into geosynchronous orbit. Truax could do it for less, way less.

In 1984 Fell Peters wrote to a marketing consultant at the University of Southern California about a strategy for attracting investors. The response he got outlined the problems facing PPE: lack of a clear timeline for expected PPE launch and development of satellite launch capability, lack of cash flow forecasts, and no assurance for nervous investors that PPE had the marketing capability to sell its launch services to prospective customers, regardless of its superior capabilities or costs.

This lack of a precise timeline and marketing capability had dogged PPE from the beginning. The complexities of the private launch industry, with its government regulations, massive funding requirements, and agency-corporate turf battles, had already undermined the efforts of other firms attempting to get a foothold in this nascent market.

When Truax was finally able to stir interest in his big dumb boosters, it came not from private industry but from the federal government. And it came less from careful business plans than from disaster. When the space shuttle began service in 1981, it had already disappointed its earlier forecasts of low cost and launch frequency. The explosion of the shuttle *Challenger* in January 1986, with the tragic loss of its crew, would completely redefine the satellite launch industry.

Suddenly, expendable launch vehicles looked good once again. Truax's proposal for rocket development, which had been circulating in various government agencies, found a receptive audience at the Naval Research Laboratory (NRL) in Washington DC. One of NRL's roles was to put up satellites for communications, navigations, or weather observation. But to launch a satellite, they had to use rockets operated by either NASA or the Air Force. How much better would it be if they had a launch system of their own? Truax's proposal to develop an inexpensive rocket to be launched and recovered in the sea was just the ticket. Thus, in 1987, the SEALAR project, Sea Launch and Recovery, came into existence. Finally, Robert Truax had the funding he had long sought to vindicate his rocket ideas.

Actually, SEALAR was simply a continuation of the Volksrocket/X-3, without the human payload and with a water launch capability. And yet everything had changed. For one, there was lots of money. Old volunteers got put on salary and new people got hired. Eventually Navy personnel also became involved in the project.

Frustrated by the long delays with PPE, Peters had married and become involved in other business ventures. But for him, the Navy contract was an abandonment of the original dream. "I told Bob, 'Finish the project, then develop this new rocket.' My heart was in making that flight, that fifteen-minute flight, making history."

All those PPE volunteers had not donated their time and talents to develop a Navy rocket, Peters argued. Teenagers from Britain had not turned over their allowance to give the U.S. Navy its own launch capability. PPE had been on a crusade to send a private astronaut into space, to hand over space access to private industry so that it could provide access more efficiently and open space to everyone. The many supporters of PPE over the years had wanted to be part of that. Now the quest to put the first private astronaut into space was being abandoned.

The SEALAR project provided Robert Truax a few flush years to explore his big dumb booster and sea launch concepts. However, Truax got caught up in the very government bureaucracy he so despised, the funding battles, turf disputes, and contractual complexities. The program did not last long enough to develop a functional rocket, and when the funding dried up, Truax got stuck with massive close-down costs for equipment and facility contracts and protracted legal battles that nearly bankrupted him.

The legacy of Truax's long career and his contrarian approach to rocket building set the standard for the swim-against-the-current entrepreneurs who would follow. Until his PPE, no one had attempted to send a private astronaut into space. That business model did not exist. Had financing materialized for that project, he might have given the space tourism business a three-decade jump-start.

Instead, the path for private citizens into space took a different course. About the time that Truax formed his engineering company in the mid-1960s, government-run space programs began to loosen their restrictions on who could fly on rockets. Average private citizens began to take their place beside the jet jockey astronauts who first showed the way.

3. Private Citizens Get Their Chance in Space

For the first time since the birth
of science, two trained observers
were rocketed into space yesterday.

Walter Sullivan, *New York Times*,
13 October 1964, commenting on the launch
of the first Voskhod spacecraft

Morning dawned chilly on 12 October 1964 on the inhospitable steppes of Soviet Kazakhstan. A bus pulled up to a launchpad at the Baikonur Cosmodrome and unloaded the men who were about to launch in the new Voskhod spacecraft. They strolled casually past the crowd of onlookers, smiling and waving, dressed in simple woolen trousers and jackets. From all appearances, they might have been embarking on a business trip on an airplane rather than on a spaceflight.

Those gathered for the occasion certainly realized what an extraordinary sight this was. For one, the men were not wearing the typical bulky space suits worn by all previous cosmonauts. More noteworthy still, there were three of them, and they were about to embark on the first multicrewed spaceflight. They were, by profession, a pilot cosmonaut, an engineer, and a physician. And that was the most remarkable thing of all.

An elevator lifted them through the rocket gantry to a platform. A short climb up a metal staircase took them to their capsule. There they removed their jackets and exchanged their boots for suede slippers, then donned communication caps. Commander Vladimir Komarov climbed aboard the capsule first, then engineer-scientist Konstantin Feoktistov squeezed in, and finally physician Boris Yegorov entered. It was a snug fit.

Voskhod was really the older, single-man Vostok capsule with a few critical safety features stripped out to make room for three passengers: no ejection seats to save them in the first twenty seconds of flight, no pressure suits to safeguard lives in the event of a pressure leak. An accelerated development program had rushed the capsule to launch prior to the pending American two-man Gemini flight.

Once in orbit, the men unstrapped their safety belts and peered out the portholes. None of them had flown before, and they were captivated by the view of the Siberian forests. Air force colonel Komarov had entered the cosmonaut program four years earlier, along with Yuri Gagarin, in the first group of pilots recruited for the Soviet space program. Feoktistov and Yegorov were not air force cosmonauts at all but civilians working in programs associated with the space program. Feoktistov designed space capsules, and Yegorov monitored the health of those who flew. This was the first time that nonprofessional astronauts had traveled into space.

Except for the special circumstance of the first woman in space, Valentina Tereshkova, all previous cosmonauts had been military jet pilots. The most telling distinction between Tereshkova and her Voskhod counterparts is that she actually entered the corps of cosmonauts and trained for sixteen months, whereas Voskhod's two civilians had trained for only three to four months.

Since 1962 the architect and "chief designer" of the Soviet space program, Sergei Korolev, had been recommending sending up nonpilots (engineers, physicians, journalists, poets), an idea firmly resisted by the Soviet air force, from which all cosmonauts were recruited. But the six Vostok flights had now proved that it was possible for a human being to fly in space, and that piloting a spacecraft required only one person. Therefore, to Korolev's way of thinking, it didn't make sense to waste Voskhod's two extra seats with additional cosmonaut pilots when the craft could carry aloft "passengers" to make observations, conduct experiments, or report on the experience.

Korolev proposed filling Voskhod's extra seats with an engineer and a physician. Months of bureaucratic wrangling followed as various governmental departments, the air force, and the Academy of Sciences argued the merits of the proposal. As was often the case, Korolev got his way.

Korolev proposed the names of fourteen engineers from his own design bureau. Eight passed medical screening, but only one (Feoktistov) was recommended for training. Of the thirty-six physicians nominated, five moved on to training.

These six candidates arrived at the Cosmonaut Training Center in early June 1964, along with a scientific researcher, proposed by the Academy of Sciences. Medical problems during training eliminated a further two candidates, leaving only five to compete for Voskhod's two available seats.

In cosmonaut selection, political qualification weighed as heavily as medical fitness and training performance. In late August, it was learned that the Academy of Sciences candidate, Georgi Katys, had a brother and sister living in Paris. They had moved there sixteen years before Katys was born. But when that fact was added to the other blot on his political record, the execution of his father by the state, he became unacceptable as a cosmonaut.

Each of the remaining four candidates had strong advocates working behind the scenes. Their infighting delayed final crew selection until early October, only weeks before launch. Korolev, who favored a crew composed of Komarov, Feoktistov, and Yegorov, again prevailed.

At twenty-eight, Boris Yegorov had been the youngest of the physician finalists. Medicine ran in the family. His mother was an oculist, his father a prominent Moscow neurosurgeon. The space age had unfolded around Yegorov. The chance reading of a journal article on the physiology of high-altitude flights convinced him to study space medicine. His education at the First Moscow Medical Institute coincided with the start of the Soviet manned space program. During his last year of school, he had been given the opportunity to watch Yuri Gagarin undergo centrifuge training. Graduation came in 1961, the year of Gagarin's historic flight.

After graduation, Yegorov served as a doctor on one of the parachute teams recovering the Vostok cosmonauts. He then moved on to the air force's Institute of Aviation and Space Medicine, where he specialized in research on the vestibular apparatus, the part of the ear that controls the sense of balance.

Konstantin Feoktistov, age thirty-eight, came with far more impressive credentials, yet it is surprising that he survived the arduous selection process. One obvious strike against him was that he was not a member of the Communist Party, usually a critical job requirement. More troubling, a host of medical problems—with his spine, his intestines, his eyesight—had led air force doctors to twice disqualify him from consideration. Nikolai Kamanin, head of the cosmonaut corps, labeled Feoktistov an "invalid."

An additional fact weighing against him was that as an engineer in Korolev's design bureau, Feoktistov had led opposition to the idea of converting the single-man Vostok capsule to the three-man Voskhod. Ironically, it would be this opposition that ultimately earned him his ticket to space.

Feoktistov joined Korolev's design bureau in late 1957 and played a key role in the design of Vostok, the first manned space capsule. Because he knew Vostok so well, Feoktistov took the lead in opposing Korolev's suggestion that the spacecraft could be converted to hold three cosmonauts. "Two or three times he [Korolev] came back to the question," Feoktistov recalled in his memoirs, "and I had to convince him that nothing good would come from failure." Feoktistov was just cantankerous enough to oppose a man of Korolev's stature and notorious temper, especially when he knew he was right. Jamming three cosmonauts into the small capsule left no room for such critical safety equipment as the ejection seats or even for space suits.

Since the roomier Soyuz capsule was then in development, Feoktistov reasoned, why roll the dice on this jury-rigged spacecraft? In light of the difficulties and dangers of converting the Vostok, it was a legitimate engineering question, but one that completely overlooked the twisted calculus of the cold war. The risk was justified in light of the American Gemini program, which planned to carry two astronauts by mid-decade. Not only would the Voskhod beat the American multicrewed flight, but it would carry not two but three crew members.

So, when Korolev once more returned to Feoktistov in February 1964 with the idea of converting the Vostok, he dangled a "hook baited with a fat worm." If the Vostok could be converted to make room for a three-man crew, one of those crew slots could go to an engineer. Feoktistov explained in his memoir, *The Trajectory of Life*, that he took this as a "gentlemen's agreement" that if he supported the creation of the Voskhod, he would be selected as that engineer crew member.

Getting into space would be the fulfillment of a childhood dream for Feoktistov, and that was one of his strongest incentives to pursue the work. However, it put him in the unenviable position of stripping safety features from the very capsule in which he himself might fly.

The Voskhod capsule may have been severely cramped for space, but no space flight had ever before had the luxury of extra passengers, with the time to observe and conduct experiments. Feoktistov spent his time in orbit testing equipment and measuring the capsule's position relative to the stars. Yegorov observed the crew's reaction to microgravity and took blood samples, blood pressure, and respiration readings.

As the second orbit brought Voskhod back over Soviet territory, Komarov radioed ground control, "Dawn, Dawn. I am Ruby. Can you hear me?"

"Ruby, I am Dawn. Reception is good."

They had achieved safe orbit, Komarov reported, and the crew was doing well. Then the Voskhod beamed to ground images of the men for Russian television. The broadcast was relayed to Western Europe via the facilities of the European Broadcasting Union (Eurovision).

The U.S. media company Columbia Broadcasting System (CBS) obtained a copy of the film in Berlin and rushed it by jet back to New York in time for the evening TV newscast. The one minute of film appeared so grainy and overcast that virtually nothing was discernible until near the end, when the shadowy outline of a face emerged. The clip provided stunning confirmation of yet another Soviet "first" in space: three cosmonauts orbiting the Earth at the same time. NASA's two-man Gemini program would not achieve its first flight until the following year.

New York Times science reporter Walter Sullivan noted an equally important first achieved by the Voskhod flight: "the transportation of specialists into realms beyond the Earth's atmosphere." Without this important objective, Sullivan commented, manned space flight would be "little more than a stunt."

As the Voskhod capsule passed over the United States on one of its orbits, the cosmonauts sent greetings from space to the industrious American people. But the Americans were picking up far more than greetings from the Soviet craft. The Foreign Technology Division (FTD) of the U.S. Air Force (USAF) was monitoring the telemetry signals of the cosmonauts' biomedical information sent to Soviet ground control. The United States had been listening in on these transmissions since the flight of Laika aboard *Sputnik 2* in 1957, allowing NASA to amass an impressive amount of medical data about the body's response to space flight.

USAF flight surgeon Duane Graveline, who worked for FTD, knew almost as much about the physiological performance of cosmonauts as did the Soviet medical staff. He had studied the medical data transmitted on the dog Laika and had been monitoring the in-flight performance of each cosmonaut since Gherman Titov's *Vostok 2* flight in August 1961.

Graveline's primary area of interest was the body's reaction to prolonged weightlessness. At the USAF School of Aerospace Medicine, he had conducted bed rest and water immersion studies of zero-g deconditioning. Combined with his dual role as an FTD medical analyst and a medical flight controller for NASA's Mercury program, this research made him one of the most knowledgeable medical professionals in the burgeoning field of bioastronautics.

So, in the same month as the *Voskhod 1* flight, when NASA announced that the agency would be recruiting its own scientist astronauts, it seemed the most natural thing to Graveline that he should apply. The same thought occurred to 1,700 other scientists, engineers, and physicians, who hurried off applications to NASA.

With longer-duration flights and lunar landings on their agenda, NASA administrators had long since recognized that science would play a larger role in future missions. Astronaut selection criteria had earlier evolved from an emphasis on military test pilots to include civilian jet pilots. NASA's third group of astronauts, recruited in 1963, had been allowed to substitute advanced academic degrees for some of the required flight hours and test pilot status. Edwin "Buzz" Aldrin, who held a PhD in astronautics, entered with this group.

Now, in its attempt to attract more scientists, NASA was opening the door wide enough to accept even nonpilots. Men who had previously not considered themselves potential space travelers suddenly saw a role for themselves in this expanded definition of astronaut.

If some NASA official had designed a career path for the ideal scientist astronaut, it would have looked something like the résumé of Dr. Duane Graveline. For him, the opportunity to join the astronaut corps was the realization of a long-held dream, the capstone to years of preparation. In that sense, he bore a striking resemblance to the "passengers" aboard *Voskhod 1*.

However, most of the men who would eventually be selected for NASA's first class of scientist astronauts had no previous involvement with the space program. For some, the dramatic accomplishments of the Soviet and U.S.

space programs influenced their career choices. Others seemed to find their way into the program almost by accident.

Among the 1,700 applications sent to NASA was one from an engineer and academic named Owen Garriott. Originally from Enid, Oklahoma, Garriott was a graduate student in electrical engineering at Stanford University when his career plans received a fortuitous boost from the launch of *Sputnik*. Many of the professors and graduate students then working in Stanford's Radio Propagation Laboratory converged on their field site to use its receivers and antennas to listen to the beep-beep-beep transmission from the Russian satellite.

Garriott had been looking for a research topic for his PhD dissertation. "What more could be provided than studying the signals from the Sputniks as they transversed . . . and had various kinds of effects superimposed upon them due to transmission through the ionosphere?" He had joined the Stanford faculty by the time Yuri Gagarin and Alan Shepard launched the era of manned space flight. A few years later, when he learned that NASA would be sending researchers along on their spaceflights, he applied for the job. To bolster his credentials, he got his pilot's license, figuring that would be something NASA would be looking for.

A geologist named Harrison Schmitt was studying in Norway on a Fulbright scholarship when the idea of space first got a grip on him. It was the fall of 1957, and again *Sputnik* played a role. More so than the technical accomplishment, Schmitt was taken by the degree to which *Sputnik* "influenced and excited and literally scared the students who were around at that time from all over the world. That really did get my attention about how important space almost certainly was going to be in the future of humankind."

A report from the National Academy of Sciences in the early 1960s suggesting that the first person on the moon should be a hard rock geologist gladdened the heart of every geologist, including Schmitt. Although that possibility intrigued him, he never thought in terms of participating himself. That changed in 1964. Schmitt was working for the U.S. Geological Survey when NASA asked for volunteers for its first group of scientist astronauts. "I thought about 10 seconds and raised my hand and volunteered. I can remember feeling, at the time, that if I didn't volunteer, no matter what happened to my application, that I'd almost certainly regret it when human beings actually went to the moon."

On 27 June 1965, when NASA announced its first scientist astronauts, the names of Graveline, Garriott, and Schmitt were among them, along with Edward G. Gibson, PhD, engineer; Joseph P. Kerwin, MD; and F. Curtis Michel, PhD, physicist. But within a week, fate dealt a crippling blow to Duane Graveline's ambition to fly in space. His wife filed for divorce. The attendant bad publicity for NASA resulted in Graveline being forced to resign from the program. The remaining candidates, who were not already qualified jet pilots, spent their first year as astronauts earning their wings.

If they were ready for space at this point, space was not quite ready for them. Former Mercury astronaut Deke Slayton, director of flight crew operations, had the decisive role in assigning astronauts to flights, and he was not all that comfortable with this new crop of astronauts.

"I didn't have anything against scientists or doctors," he conceded, "but I wasn't quite sure what I was supposed to do with them on flight crews." With the Apollo crew limited to only three astronauts, scientists were simply not the first priority. "[It] takes two or three people to make sure the space craft gets where it's supposed to go, makes all the planned maneuvers, then returns safely." If a crew had problems, what they needed was someone as qualified as they were to fly the spacecraft, leaving no room for what would basically be a passenger.

Despite Slayton's concerns, Jack Schmitt did get to fly on Apollo, becoming the first geologist on the moon in December 1972. Apollo and the work of space scientists came into finer alignment when the Apollo Applications program led to the orbital space station *Skylab*, on which scientific work could be performed. Aside from Graveline and Curtis Michel, who resigned in 1969 to resume his science career, all of the rest of NASA's first scientist astronauts flew on *Skylab* missions during 1973–74.

The advent of space stations, the U.S. *Skylab* and Soviet *Salyut*, dramatically increased the number of crew slots available to scientists. NASA coined a new name for these astronauts — mission specialists. They were fully trained astronauts, given special assignments for a particular mission.

Over the next decade, as the Apollo program wound down and focus shifted to the Space Transportation System, otherwise known as the space shuttle, the role of the space traveler would once again be redefined. Deke Slayton's claim about needing everyone onboard a space craft to fly it was no longer the case. As space stations appeared and multicrewed spacecraft entered service, the science agenda could be more fully implemented and commercial and political agendas would also begin to influence crew selection.

In the 1970s, as the space shuttle emerged as the next-generation space vehicle in the U.S. fleet, crew selection would be driven by its capabilities. The shuttle was expected to fly every few weeks. With a crew capacity of seven and an ambitious schedule of scientific and commercial missions, it would need a lot of astronauts.

In January 1978 NASA announced the names of thirty-five new astronauts for its space shuttle program, its largest class ever. Twenty of this group fell into a new category of scientist astronaut, the mission specialist. This influx of scientists brought a clash of cultures with the military pilots who had traditionally made up the astronaut corps.

Air Force pilot Mike Mullane entered the astronaut corps in the January 1978 class. In his hard-hitting book *Riding Rockets*, he mentions the tensions between pilot astronauts, who tended to be conservative, military veterans, and the new crop of scientists. "These were men and women who, until a few weeks ago, had been stargazing on mountaintop observatories and whose greatest fear had been an A– on a research paper. Their lives were light-years apart from those of the military men in the group. . . . We were test pilots and test engineers. In our work a mistake wasn't noted by a professor in the margin of a thesis, but instead brought instant death."

However, the training regimen for the Thirty-Five New Guys (TFNG), as this new class was called, was built around not creating any distinction between individuals. Mission specialist James "Ox" van Hoften remembers that first year of training as a "big party," with a lot of opportunity to build camaraderie. They traveled to some of the NASA centers and visited contractors and the U.S. Congress, got to climb on a lot of hardware and be part of the NASA public relations effort.

Meanwhile, the only big issues simmering beneath the surface were how a candidate could stand out from the crowd and then be selected for a flight. Van Hoften, a naval aviator with a PhD in hydraulic engineering, recalled sitting through endless classes without ever being tested to see if he had learned anything. Even that traditional method of ranking success was not employed. "So it lent an air of real mystery, though, as this went on, about how do you get selected for a flight. That's the big question *du jour* in NASA, and no one knew."

As training continued, the candidates were assigned to work with some of the more senior astronauts—Ken Mattingly, John Young, Bob Crippen. Van Hoften put in a stint with NASA contractor Rockwell International in Downey, California, where he spent twelve hours a day in a simulator verifying reentry software. More enjoyable duty came in the form of working astronaut support at Kennedy Space Center. But, according to van Hoften, the burning question never went away. "After about a year in the program everyone just wanted to know how they were going to get in line to fly. That's all there was to it."

The hardest part of the process came with the crew assignments for the early shuttle flights (STS-7, -8, and -9). Those whose name did not appear on the list were left to wonder what they would be doing and where they stood in line. But the wait in line was about to get longer for some of the TFNGS as yet another category of astronaut stepped into the picture to capture some of the coveted flight assignments.

Much of the science planned for the space shuttle would take place in Spacelab, science modules that could be configured to meet specific mission requirements. Under a memorandum of understanding with NASA, Spacelab was being built by the European Space Agency (ESA). The first Spacelab module would be given to NASA in exchange for flight opportunities for European astronauts. Prior to NASA's announcement of its January 1978 group of astronauts, ESA had already announced its own candidates for a future Spacelab mission. Along with MIT researcher Byron Lichtenberg, four European candidates would join the NASA roster in May 1978. Given the title of "payload specialist (PS)," they had been selected by the scientific organizations to operate experiments traveling aboard Spacelab.

Payload specialists would be the first nonprofessional NASA astronauts to train for a spaceflight. Their training would be far shorter and far less thorough than that of an astronaut. Typically lasting for three to four months, the abbreviated training allowed them to gain a basic understanding of shuttle systems and how their experiment would operate on board.

One would-be astronaut who did not get selected in the shuttle 1978 astronaut class was an astronautical engineer named Charles Walker. Walker had been bitten by the space bug while still in high school. He hailed from Indiana, about twelve miles from the hometown of Mercury astronaut Virgil "Gus" Grissom. Grissom's Mercury and Gemini missions planted the idea in Walker's head that even a small-town boy from Indiana could get into space.

However, unlike the Mercury astronauts, Walker was not a military test pilot or a perfect physical specimen. So he figured his role in space would be to design spacecraft rather than to fly in them. He earned a degree in astronautical engineering from Purdue University, Grissom's alma mater, and then worked at Bendix Aerospace Company and the Naval Sea Systems Command.

In the long run-up to the shuttle program, however, Walker's thinking began to change. "It was only in the 1970s, as NASA began to build the first reusable spaceship, that I came to the slow realization that, 'It's not just the test pilots, not just the Bob Crippens of the world, that are going to get to go into space, but scientists and engineers too. Gosh, I am an engineer. Maybe I can do this.'"

Encouraged by the knowledge that NASA would be recruiting scientists and engineers as nonprofessional astronauts this time around, Walker submitted his application to become a shuttle astronaut. He didn't get very far. Lacking both an affiliation with a major university and a PhD doomed his candidacy.

Before NASA even announced its new class of astronauts in January 1978, Walker had already formulated another plan for getting into space. There had been talk of payload specialists being used to support possible private-sector experiments that could fly on future shuttle missions. Walker saw this as a possible avenue to space. "That's what I literally had in mind as one part of my professional pursuit when I was interviewing with McDonnell Douglas and the other aerospace companies in 1977, looking for that project that might actually produce in their design and development something that NASA would be willing to fly and that eventually would be of such interest and maybe practical use that there would be the real need for a payload specialist researcher to go along with it."

Luck went his way in 1979, when he took a position with McDonnell Douglas Astronautics Company (MDAC) — the one firm that was working on a commercial project for the shuttle. MDAC had a joint endeavor agreement with NASA to fly a continuous flow electrophoresis system (CFES). Electrophoresis is a chemical process by which compounds are separated out of a fluid under an electrical charge. Operated in the microgravity of space, it had the potential to purify medical compounds for pharmaceutical use.

At MDAC, Walker quickly ended up as test engineer on a small team developing the CFES. "I basically volunteered right off the bat in '79," he recalls, "and again, my management knew . . . that I wanted to be the first to have the opportunity to go fly, if somebody from the private side should have the opportunity to fall into one of those payload specialist positions."

Preparation of the project progressed during 1980 and 1981, as the orbital flight tests of the space shuttle got under way and the crews were designated for upcoming missions. When CFES got assigned to STS-4, interaction between MDAC and NASA intensified. Astronaut pilot Henry "Hank" Hartsfield was assigned to operate the CFES device during the mission. Walker trained him during the summer of 1982, using a high-end CFES simulator.

Some of the complexities of the operation became apparent on the maiden flight of the CFES on STS-4. This was prior to the existence of the Tracking and Data Relay Satellite, so the crew was out of touch with ground control for a good portion of every orbit. During the mission, Walker sat in the back room at the mission operations control room at JSC. "There was plenty of time in which you're sitting back there wondering, 'What's going on now?' and you can hardly wait until acquisition of signal."

The CFES was a sophisticated piece of machinery that required careful monitoring and occasional adjustments to valves, flow rates, pump speeds, and electric field to stay within critical parameters. But CFES did not connect to the onboard telemetry system. Instead of receiving a continuous flow of data, they got information only when an astronaut could read out numbers from the machine and transmit them to ground. As Walker related, sometimes the pass was so short, and there was so much else going on, that "the crew would just give a quick verbal burst of information. 'I've gotten one point three point two, Charlie Oscar, three five.'" Walker and the MDAC team then pored over those numbers, trying to figure out what was happening and what to do next.

The whole process was additionally complicated by the fact that Walker and his team could not talk directly to the crew; only the CapCom (capsule communicator, always a fellow astronaut) could do that. Any question had to go through the payload officer, who took it to the flight director, who would then okay it for CapCom to put to the crew. This less than perfect arrangement prompted MDAC to make a proposal that would add yet another dimension to passengers aboard the shuttle.

In a 2004 NASA oral history interview, Walker recalled that very important moment when he was handed his ticket to space. He sat in a meeting with Jim Rose, head of the CFES team for MDAC, and Glynn Lunney, NASA's manager of the Shuttle Payload Integration and Development program. Rose conceded that the astronauts were doing alright tending the CFES when they had time to do so. But MDAC would get better results if there was a payload specialist devoted exclusively to the CFES, someone to monitor it nonstop and make all the necessary adjustments.

The proposition sat in the air for a moment as Lunney chewed thoughtfully on a cigar. It wasn't a totally foreign idea, since payload specialists were already in training for Spacelab. Then Lunney asked Rose if he had anyone particular in mind. Rose turned to Walker and said, "You're looking at him."

It all made perfect sense that the guy who helped to develop the machinery and train the astronauts to run the experiment would travel along with the CFES to operate things himself. It required only a small alteration to the NASA-MDAC joint endeavor agreement. The agreement covered flying the CFES equipment but not a payload specialist to accompany it. MDAC would have to pay for Walker's flight. That had never been done before. How did you calculate the cost of sending someone into space, the cost of adding just one person to a mission already set to fly? No one had ever paid for a flight into space. The figure NASA eventually came up with was $40,000.

Following an application process, Walker was added to the crew list for STS-41-D, a future mission on which CFES was already manifested. Within days of receiving notice in May 1983, he traveled to Houston to meet the crew and trainers. In one respect, Walker's training would be easy. He was already familiar with many of the astronauts and with astronaut operations at JSC. CFES had already flown on three missions, and Walker had trained the astronauts who operated the machine for those flights. In fact, commander of STS-41-D Henry Hartsfield Jr. had operated the CFES on its maiden flight aboard STS-4.

Walker's training began in the summer of 1983, about a year before his scheduled launch date. It would create the model for another category of shuttle passenger. Planning for the flight of NASA's first commercial payload specialist just happened to correspond with its plans to move in another bold new direction, the flying of civilian passengers. Back in 1982, NASA administrator James M. Beggs had requested a study of the feasibility of allowing private citizens to fly on the shuttle. The report from the Task Force for the Study of Private Citizens on the Shuttle appeared in the same month that Walker began training for his flight into space.

As Beggs explained it at the time, "The astronauts all come back from these flights and they all say, 'Gee whiz, we have these pictures and they look beautiful but they don't near do justice to what we saw.' So we'd kind of like to send people up there who can translate the experience of what they see in space into real terms for the public. That implies artists and journalists and writers and what have you."

Flying private citizens was a natural outgrowth of the shuttle vision that had been sold to Congress in the 1970s. The shuttle would fly with commuter-like regularity, projected at twenty-five to seventy flights per year. By the summer of 1983 there had been only seven shuttle flights in the first two and a half years of operations, but the pace was about to accelerate, and there would be opportunity now to deliver on that promise.

When Walker began his training, he pushed into the unfamiliar territory of a civilian in the ranks of the astronauts. Scientists had traveled this path before him, and NASA had other payload specialists in training for Spacelab, but Walker would be different. He would remain an MDAC employee, splitting his time between company headquarters in St. Louis and the Johnson Space Center in Houston. "Working out the training schedule was interesting," he recalled in a 2005 interview. "I was not going to be a resident at JSC or in the area. I literally was an itinerant. Coming and going was the arrangement, and it was decided that that should be sufficient."

The other members of the 41-D crew had been training for several months when Walker joined them in the summer of 1983. Launch was expected in spring of 1984. The training syllabus for the other payload specialists served as a template, but Walker's training evolved as it went along. "We were working it out in real time as we went through, and it was kind of an ongoing negotiation between myself, my employer, the McDonnell Douglas Astronautics Company, and NASA, JSC and the crew, the Flight Crew Office, MOD [Mission Operations Directorate], and the schedulers as to just what my syllabus and what my schedule would be."

Basically a familiarization training, Walker's preparation included spaceflight orientation, orbiter system orientation, shuttle operations and habitability instructions, and safety procedures training. One backseat flight in a T-38 jet trainer introduced him to g forces, and about forty parabolas on the KC-135 aircraft, NASA's "vomit comet," provided a good introduction to weightlessness and the space adaptation syndrome he would actually experience aboard STS-41-D.

Study of textbooks and training materials was interspersed with time in the classroom, one-on-one training with an instructor, and computer simulations. "It was finally decided by NASA for me and my circumstance, the training was basically familiarizing with those systems 'so he knows what's going on around him, but he's not going to operate any of those.'"

Familiarization also extended to learning how to work with the crew. He would literally look over their shoulders, watching as they learned about the orbiter's systems. Since he had a limited amount of time to train with them, he was also invited to numerous social occasions in the homes of the crew members. "The part that really took the most time was the training to become one of each of the crews that I flew with; in other words, the crews getting to know me, and me getting to know my fellow crewmates [so that] we could work together and feel good working together and flying together as a team."

One month prior to launch, Walker took up residence in Houston and remained there full-time for completion of training and final preparations for launch. Preparing this civilian for space was as much a learning experience for NASA as the flight preparation was for Walker. One of the lessons learned from his own training, according to Walker, "was that those guys actually can be trained in just a few months' time . . . and become a good, acceptable working complement to a crew."

The need to define appropriate flight training for civilians was driven home on 27 August 1984, just three days before Walker's first trip into space, when President Ronald Reagan announced, "I am directing NASA to begin a search in all of our elementary and secondary schools and to choose as the first citizen passenger in the history of our space program one of America's finest: a teacher."

NASA's motivation to send civilians to space came as an outgrowth of an explosion of public interest in the space program, thanks to the shuttle. Its main purpose was to maintain and build that interest. An informal NASA task force studying the issue in 1976 had identified the types of individuals who should be considered for such a mission. They included journalists, politicians, prominent scientists, artists, humanitarians, entertainers, and "laymen."

When the shuttle actually began to fly, letters poured into NASA from citizens eager to go into space. One common refrain was "When is a normal taxpayer going to get to fly?" Maybe it was too late for all those people who had written to Robert Goddard asking to fly in his rocket, but still waiting in the wings were the more than eighty thousand people who had joined Pan American Air Line's First Moon Flights Club in the 1960s. Some of the citizens writing to NASA when the shuttle went into service actually sent along copies of their First Moon Flights Club card with the question, "When can I turn it in?" Obviously, the agency would need a procedure for determining who should get one of the coveted seats aboard the shuttle. Another NASA task force took up the challenge.

On 21 June 1983 the Informal Task Force for the Study of Issues in Selecting Private Citizens for Space Shuttle Flight sent its final report to NASA administrator James Beggs. "Flight of private citizens is both feasible and desirable," it stated, "provided that their flight be for purposes included within the scope of the Space Act."

Because the Space Act that created NASA charged it to "provide for the widest practicable and appropriate dissemination of information concerning its activities," the task force recommended that citizens be drawn from the ranks of written communicators, visual communicators, and teachers. At first NASA thought it would simply pick someone who could best represent the space experience to the public.

According to Alan Ladwig, who would eventually head up the NASA Space Flight Participant program, "one of the suggestions was Philippe Cousteau, the son of Jacques Cousteau, with the thinking that he could do for space what Jacques Cousteau had done for oceanography. And if not Philippe Cousteau, what about Walter Cronkite, because he was so affiliated with covering space?" However, it was finally decided that there would have to be some kind of selection process.

Ladwig served on the NASA committee formed to establish the process for selecting a private citizen to fly on the shuttle, which would later take the official designation of space flight participant. Brian Duff of the National Air and Space Museum had already highlighted the importance of this development. Writing in *Newsweek*, he claimed that the participation of citizens in the shuttle program would have a significant impact on our culture. "It would help to reinforce the belief that this country in the 20th century is a space faring nation."

In late April 1984 the committee submitted its recommendation to Beggs that a teacher be designated as the first apace flight participant. "He took a long time to respond to the recommendation," Ladwig recalled in a 2008 interview. After all, flying a private citizen represented a historic turning point in NASA's mission.

It was not until late June that Beggs finally issued a memo approving the committee's recommendation. "He wrote at the bottom of the memo, 'longest day of the year,'" Ladwig recalled. "For the longest time I thought he wrote that because making that decision had been of such importance and was so wrenching for him. It wasn't until several years later that someone pointed out to me that the date on the memo was June 21, which *is* the longest day of the year."

Even if that notation did not signal the importance of the event, Ladwig stated, "The political appointees in the Agency decided that this was of such historical significance that the President should make the announcement." It was all very hush-hush so that President Reagan could break the news. "At this time *Life* magazine was working on an article speculating about what private citizen would be selected. They already had a layout that included photographs of movie stars." Ladwig, who was sworn to secrecy, phoned the magazine and informed them as carefully as he could that it might not be a movie star flying on the shuttle but a more ordinary citizen, maybe even a teacher. The message got through, and the photograph of a teacher appeared with the article.

President Ronald Reagan made the announcement on 27 August 1984 that a teacher would travel on the shuttle. Within days of Reagan's announcement, teachers began writing to inquire and volunteer. "Look no further; I'm the one," was the most frequent statement in the letters.

Astronaut David Leestma would later describe 1985 as the "hoo-ha, 'Let's go fly' year," as if the groundwork had been laid and NASA could now pull out all the stops. The year when the shuttle would finally begin to realize its promise would see more flights (nine), larger crews (seven became the norm), international payload specialists, the flight of a senator and a congressman, and the creation of the Teacher in Space and Journalist in Space programs.

Despite Beggs's endorsement of the Teacher in Space program and Reagan's media-grabbing announcement, Ladwig found that he had to champion the program within the agency. "The attitude at JSC was, 'Do you really expect us to fly a teacher?'" Resistance to shuttle "passengers" also reached its peak among the astronaut corps about this time.

Mission specialist Mike Mullane leveled the harshest criticism at the Space Flight Participant program, calling it "immoral." This opinion held that since the shuttle was still too dangerous, it was immoral to fly passengers for what amounted to public relations purposes.

The more widely shared criticism of the spaceflight participants was simply that they were even more bodies getting in line in front of the career astronauts. By this point nonprofessional astronauts had flooded the ranks. Not only the spaceflight participants but corporate payload specialists, scientists from Europe and Canada, military space engineers flying secret missions, foreign citizens offered a free ride if their countries launched a satellite on the shuttle, and politicians all clamored for a place.

They were stealing seats from career astronauts, more than twenty of them in the shuttle's first five years of operations. These spots would otherwise have been assigned to mission specialists, who had trained long and hard and waited patiently for their mission slot, only to see their seat go to someone who came and went from the program in the course of a few months. Bad experiences with payload specialists compounded the bitterness. On rare occasions, crews had to nursemaid sick or emotionally unstable payload specialists, sometimes to the point of concern for mission safety.

The way Ladwig saw it, this attitude repeated an old pattern. "Each time the door to space has opened a little wider, it's been met with resistance, from the first non-test pilot astronauts, the scientists, through the mission specialists, payload specialists, and spaceflight participants."

Since the Teacher in Space program would lay the foundation for other participant programs to come, it was important that a fair selection procedure be developed using appropriate professional standards. NASA chose the Council of Chief State School Officers (CCSSO) to help develop the application and coordinate the teacher selection process. The council's daunting fifteen-page application would be the first hurdle faced by would be teacher astronauts. However, teachers were inured to filling out such forms: tens of thousands of them requested the application.

Ladwig served as the public face of the program, working with the media to build interest. On 7 November 1984 he appeared on the David Letterman TV show to discuss the Teacher in Space program. With the advent of the shuttle, a generation of young people would grow up in an age when space travel would be more commonplace, he explained. As the first private citizen in space, the selected teacher would play a vital role of engaging the interest of young people in space and raising their appreciation for the importance of the nation's space program.

The very next day Ladwig learned the shocking news that U.S. senator Jake Garn would become the first private citizen in space rather than the yet-to-be-chosen teacher. Garn, who chaired the appropriations subcommittee that oversaw NASA, had been invited by NASA administrator James Beggs to fly aboard a shuttle mission. It was "appropriate for those with Congressional oversight to have flight opportunities," Beggs explained, "to gain a personal awareness and familiarity." Although Ladwig handled the Space Flight Participant program, he had not been aware of the Garn offer.

Garn, a former Navy pilot, had long made known his desire to fly on the shuttle. At a May 1981 hearing of his Senate subcommittee, he posed the question, "When do I get to go on the Space Shuttle?" Especially for a project as expensive as the shuttle, he would later assert, it was his obligation to "kick the tires" on a shuttle, just as he had test-flown the B-1 bomber prototype and taken the Army's new battle tank for a test drive.

Given that NASA would be going to Congress in 1985 to request $8 billon for its space station development program, perhaps it was politic to honor his request. It wouldn't hurt to have someone on the committee who had actually flown in space.

In this "hoo-ha" year, the shuttle flights of two politicians would play out against the high drama of the selection and training of the teacher in space. Garn would be launched aboard the space shuttle *Discovery* on 12–19 April 1985, before the first teacher to fly in space had even been selected. Florida congressman Bill Nelson, whose congressional district included the Kennedy Space Center and who was chair of an important House subcommittee, had been equally aggressive in seeking a shuttle flight. He would get to travel on STS-61C on 12–18 January 1986, two weeks before the teacher in space mission.

Preflight, a *New York Times* editorial had called Garn's proposed flight the "ultimate junket," while some of the crew members had bristled at the distraction of having him onboard. Mission specialist Mike Mullane groused about payload specialists in general, but reserved his strongest venom for the politicians. How ridiculous was it to think they could gain a better understanding of NASA operations by taking a ride on the shuttle? Rather than doing the hard work of going behind the scenes to see the competing pressures of system development, budget, scheduling, and safety, they opted for the more glamorous alternative of a shuttle ride.

Before Garn arrived at JSC for his two months of training, some jokester posted a sign-up sheet on the bulletin board in the astronaut office. The sheet asked for volunteers who wanted to take an eight-week course to become a senator. There was some smug satisfaction in the NASA corridors when it became known that Garn had been almost incapacitated for several days of his flight with the debilitating middle ear syndrome known as space adaptation sickness. He was even more of a passenger on the shuttle than the astronauts had predicted.

Perhaps Congressman Don Fuqua summed up best one school of thought on civilian missions in general, especially those by his political colleagues. Because of his chairmanship of the House Committee on Science and Technology, Fuqua would also have been eligible for a shuttle flight, but he adamantly opposed the idea. In a 1999 NASA oral history interview, he recalled his conversation with James Beggs:

I said, "Jim, I don't want to go. Even if you told me I could go, I don't want to go. And I'll tell you why. I don't think I have any business being there. This is still a very hostile environment. I'm not even sure you ought to have the Teacher in Space Program going. But I've got a full-time job being a member of Congress and these other guys do, too. And I don't think you've got the time to devote to proper training when you've got a whole cadre of astronauts that are professionals in this business, mission specialists, and so forth; and I don't think I ought to bump somebody to go take a ride in space. I would love to! I'd give my right arm to go, but I don't think I have any business doing that.

A large number of teachers did not share Fuqua's opinion. Of the 47,000 who requested the application, 11,400 had applied by the February 1985 deadline. Through a series of state competitions, the CCSSO narrowed that field to a manageable 114.

Huge media attention accompanied the process in each state. Applicants were featured in hometown newspapers and on local and national TV. The 114 state finalists went to Washington DC in June 1985 for a five-day national awards conference. Before leaving home, they had each sent a videotaped interview to be previewed by the selection committee.

Beggs appointed them all space ambassadors for their states. They would take a pro-space message back to their schools and communities. But the purpose of the gathering was to narrow the field to just ten finalists. The astronauts had bought into the process by this point. Gene Cernan, Ed Gibson, Harrison Schmitt, and "Deke" Slayton served among the judges in the selection process.

The ten finalists, announced on 1 July, then flew to JSC in Houston for a week of medical and psychological screening, then back to Washington. With President Reagan in the hospital, Vice President Bush would do the honors of announcing the winner. NASA received strict orders not to tell the finalists who had been selected. That announcement would only come at the official ceremony in the Roosevelt Room of the White House.

However, the finalists would have none of that, Ladwig recalled. "They didn't want it to be like a beauty contest. So we told them. When Christa McAuliffe's name was read out by the Vice President, she was still overwhelmed and began to cry."

As if to dampen the blow of Senator Garn having already flown three months earlier, the vice president mentioned that McAuliffe was the first person with "no connection to the space industry" to be chosen to fly on the shuttle, the first "ordinary citizen." Over the next six months she transformed from an ordinary citizen to both a citizen astronaut and a celebrity.

When McAuliffe applied for the position, she knew that being a public figure was part of the bargain. The Teacher in Space program was planned as a publicity vehicle for NASA, and it didn't disappoint. Photogenic and at ease with the media, public relations came naturally to McAuliffe. Her appearances on popular television shows and in interviews by the likes of Larry King, Johnny Carson, David Letterman, and Regis Philbin made her the public face of the space program and an ambassador for the teaching profession.

Although her role forced her to live and work far from home and her family, she tried to preserve her private life and a semblance of normalcy. To shield her family from the spotlight, she refused to let NASA send a film crew home with her during training breaks at Thanksgiving and Christmas. They wanted to film her enjoying the holidays with her family, but she demanded to keep that part of her life separate.

Three months of training began for her at the JSC in Houston in September. One of the first things that Commander Dick Scobee said to her and backup Barbara Morgan when they first met was "Do you really understand how risky this is?" They did, of course, and both threw themselves into the training to prove they could work just as hard as the rest of the crew.

Scobee knew the real benefits deriving from flying teachers. During their training, he told a journalist, "The short-term gain is a publicity gain. The long-term gain is getting expectations of the young people in this country to the point where they expect to fly in space, they expect to go there, they expect this country to pursue a program that allows it to be in space permanently to work and live there, to explore the planets."

From experience with payload specialists and Senator Garn, JSC had the astronaut short course nailed down: 114 hours of familiarization, T-38 flights for a taste of g forces, parabolas on the KC-135 for some weightlessness and the attendant nausea. Simulation training mainly covered living and working in the shuttle, as well as evacuation procedures. They also learned how to eat and sleep in weightlessness, and use the tricky shuttle vacuum toilet.

What couldn't be scripted is how well the nonprofessional flyer would mesh with the astronaut crew. Joint training sessions provided opportunities for team building, and social occasions helped form the bond, with varying degrees of success.

The frequent flights of 1985 and the occasional reshuffling of missions added to this challenge. The stress of dealing with last-minute changes could be compounded when new payload specialists got tossed into the mix. Such was the case when thirty days prior to the launch of STS 51-D it became 51-G.

Mission specialist Steven Nagel remembers the tensions of the switchover when two commercial payload specialists, Charlie Walker and Greg Jarvis, were replaced with Saudi Arabian Prince Sultan Al-Saud and Frenchman Patrick Baudry. "Here it is March. We're going to fly in June," Nagel recalls. "We've got different satellites. We've got all this other stuff to train for, and now we've got two other PSS, and one is from Saudi Arabia."

The crew figured the Frenchman would work out well. Baudry had trained for two years with the French space agency CNES (Centre National d'Études Spatiales) and at Star City as backup crew member for a French-Soviet *Soyuz T-6* mission. Because the flight would now carry a satellite for the Arab Satellite Communications Organization (Arabsat), Saudi Arabia was offered a PS spot. Al-Saud, a member of the royal family, a Saudi air force pilot, and fluent English speaker, got the nod.

Mission pilot John Creighton worried about the cultural gap between the crew and Al-Saud, so he arranged for someone from the Houston oil firm ARAMCO (Arab American Oil Company) to provide the crew with some cultural awareness training. Nagel remembers them as very good briefings, but the thing that stood out the most for him was "No camel jokes. No harem jokes. Don't do that around them."

In this way, the crew was able to take with them a heightened cultural sensitivity for their meeting with Prince Sultan bin Salman bin Abdulaziz Al Saud. The first thing the prince told them was a camel joke and a harem joke. This mass communications graduate of the University of Denver hit it off well with the crew. During the mission, he performed experiments designed by Saudi scientists and conducted some Earth observations. Still, it was an example of the added distraction sometimes offered by the nonprofessional astronauts.

The flights of politicians Garn and Nelson were almost universally seen as a thinly disguised attempt to curry favor with those who held the purse strings. Beggs had to extract a promise that they would devote sufficient time in their busy legislative schedules to actually complete two months of training. Both brought a legislative aid with them to JSC to coordinate their two lives during training, and they would occasionally leave training to catch important votes in Congress.

While most shuttle passengers were expected to basically stay out of the way, Nelson pushed to be given some significant role in the mission. His most troubling request was to have NASA arrange for him to talk in orbit to the cosmonauts aboard the Russian *Salyut* space station. The technical and political complexities of such a request would have been a major distraction. The crew wanted no part of it, and so it was not pursued. Garn became memorable for his prolonged bout of space sickness during the mission, which gave credence to the charge of critics that nonprofessionals did not get enough training.

Unlike her political counterparts, McAuliffe had actually put her professional life on hold for a year and committed to the mission and the space program. It made for an easier connection with the crew. In a prelaunch interview, she praised her crewmates as "very concerned about how I am fitting in with them, trying to make me part of the team." The question of how well she was able to mesh with the astronauts on her crew was illustrated during an earlier portrait session in Houston. Her crewmates told her to wait alone in the adjoining room. Although puzzled, she complied. They soon joined her, each wearing a mortarboard and holding a plastic lunchbox and an apple.

In that same prelaunch interview, McAuliffe mentioned how delighted she was that a teacher had been chosen as the first spaceflight participant. She encouraged others to follow in her footsteps. "I'm hoping that everybody out there who decides to go for it—the journalist in space, the poet in space—whatever the other categories, that you push yourself to get the application in."

Her remarks came barely a week after the deadline for applications for the Journalist in Space program. The opportunity had drawn a response from 1,700 applicants, including such well-known figures as NBC anchorman Tom Brokaw, *The Right Stuff* author Tom Wolfe, ABC White House correspondent Sam Donaldson, and former CBS anchorman and veteran space reporter Walter Cronkite. The new, less stringent medical requirements that NASA had instituted for civilian flights were forgiving enough to allow even sixty-nine-year-old Cronkite to have been considered.

The lower medical standards used for screening civilian space passengers required only that they be free of disease and injury or other condition likely to interfere with the mission or preflight training, have eyesight correctible to at least 20/40 in the better eye, be able to hear a whispered voice from three feet away (hearing aids were permissible), and have a blood pressure reading of less than 160/100. Cronkite joked at the time that "there ought to be a great advantage to prove that any old fart can do it."

With the close of applications for journalists, Ladwig had already taken preliminary steps for an Artist in Space program. It had not yet been approved by NASA, but he had called the National Endowment for the Arts to discuss its possible role in the selection process.

Several postponements of the launch of the Teacher in Space flight frustrated the crew and those who had gathered to watch. Launch was to come on Saturday, 24 January 1986, everything being timed for the key audience of schoolchildren to watch from their home or classrooms the launch of their space teacher and later catch the lessons planned in flight. However, dust storms in an emergency landing area pushed the launch to Sunday, then a shower at the Cape delayed it to Monday, and high winds caused a third postponement.

Ladwig had waited through the delays. He desperately wanted to see the culmination of the program with McAuliffe's launch. But flying a teacher was just the start of the Space Flight Participant program. Coming years would likely see two or three citizens per year flying on the shuttle. In fact, calling him back to Washington even now was a meeting about the Journalist in Space program.

On Tuesday, 28 January, Ladwig watched the launch countdown progress at his office before heading into a meeting. The Association of Schools of Journalism and Mass Communication had taken on the task of working through the 1,700 applications, under the direction of the College of Journalism at the University of South Carolina.

Within three months, the number of applicants had to be reduced to one hundred regional semifinalists. Following a procedure similar to the teacher selection, forty would be chosen as national semifinalists by May, and then that number would be reduced to ten before a winning candidate was selected.

Someone interrupted the meeting to announce that the *Challenger* had just lifted off the pad and everything was go. Ladwig felt a sense of relief. Now that NASA's first citizen in space was a reality, full attention could shift to the second. With four months of required training, and a planned September launch, the pace was about to pick up. Discussion had no sooner resumed in the meeting then a second interruption brought the terrible news that the *Challenger* and its crew were lost in an explosion.

That night, President Ronald Reagan had planned to laud the Teacher in Space program during his State of the Union address, saying, "Tonight while I am speaking to you, a young secondary school teacher from Concord, New Hampshire, is taking us all on the ultimate field trip, as she orbits the earth as the first citizen passenger on the Space Shuttle."

Within hours of the tragedy, however, a different scenario unfolded. Reagan addressed a shocked nation, consoling it for its loss and reassuring it that "there would be more shuttle flights and more shuttle crews and, yes, more volunteers, more civilians, more teachers in space. Nothing ends here; our hopes and our journeys continue."

However, the persistence of the Reagan administration's civilians-in-space vision bumped into hard reality in the aftermath of the disaster. Until the engineering and organizational problems that caused the *Challenger* explosion were identified and corrected, the shuttle did not fly. When it finally did resume service, more than two and a half years later, NASA was a different entity and the shuttle a different program.

It was no longer seen as a reliable, cost-effective space truck, catering to corporate customers and civilian passengers. The agendas of science and exploration replaced those of commerce and exploitation. Gone too was the Space Flight Participant program and with it any immediate thought of putting private citizens into space. In the coming years, the flying of civilians would once more return to where it had begun — the Soviet Union.

4. Russia Commercializes Space

If you wanted to work for the socialists,
you worked for NASA; if you wanted to work
for the capitalists, you worked for the Russians.

Jeffrey Manber

In May 1990 members of the Houston Space Society gathered at the home of Jim Davidson in Friendswood, Texas, to put together the society's monthly newsletter, *The Colonist*. It was the dawn of a new decade. NASA had emerged from its post-*Challenger* period of questioning and internal scrutiny to resume regular shuttle flights, launching satellites, the Hubble Space Telescope, and planetary probes.

But for space enthusiasts like these, who had cut their teeth on L5 Society projections of colonies in space, NASA had clearly lost its way. There was no longer a great push to put private citizens into space. The teachers, journalists, artists, and politicians who flew, or stood in line to fly, had disappeared from the scene. One tragic mishap and NASA shut the door on that great adventure, dashing the hopes of the millions who shared that dream.

Nearby Johnson Space Center ran the one NASA project with the potential to put more people in space, the space station *Freedom*. When first conceived in 1984, the station was to be an orbital outpost for science and commerce, as well as a way station for missions to Mars. But years of design and budget problems had culminated in a devastating report that uncovered serious design flaws that threatened to scuttle the project.

The time seemed right for bold considerations.

"What is the one thing the Houston Space Society could do in the next ten years," society president Howard Stringer asked, "that would change the way people think about space more than anything else?"

"We could put one of our members into space," came Davidson's quick response.

In the ordinary scheme of things, the suggestion would have seemed ludicrous, since NASA was out of the civilian-flying business. But a happy and surprising confluence of events had unfolded over the past few years to give the idea some credibility. The Soviet Union had entered the commercial space business.

When the *Challenger* tragedy grounded the U.S. shuttle fleet in 1986, NASA ceased to be the world's principal launcher of commercial satellites. Arianespace, the French-led European space consortium, was quickly overbooked with launch contracts, and an unlikely new competitor, the Soviet Union, took the first tentative steps toward commercializing its space program by offering the powerful Proton rocket as a launch vehicle.

In part, the move was born of desperation. The sweeping political and economic restructuring implemented by the new Soviet leader, Mikhail Gorbachev, had introduced free-market forces into the Soviet economy. Industries that had previously enjoyed lavish state subsidies, such as its space program, saw dramatic budget cuts. They were now expected to be self-financing, operating much as free-market enterprises, covering expenses with revenues. The end result was that the Soviet space program, which had been shrouded in great secrecy, suddenly emblazoned its top-secret Proton rocket on slick brochures and hired agents to promote its satellite launch services.

In the United States, the Reagan administration supported the same impulse toward greater commercial development of space. In fact, it was the Department of Commerce, headed by Malcolm Baldrige, that systematically resisted NASA's monopoly on space activities. In the final months of the Reagan presidency, the department created the Office of Space Commerce (OSC) to advocate for the interests of private industry in space.

Space journalist Jeffrey Manber had been writing about commercial space at that time and was asked to join the OSC. He would be on hand when the office tried to get the shuttle out of the satellite launch business. There were aerospace companies that could do the job and for much less money, as could the Russians. Unfortunately for the Russians, their launch services ran up against the U.S. Arms Export Control Act and its International Traffic in Arms Regulations (ITAR), which controlled the export of defense-related articles and services. ITAR prevented communications satellites from being sent to Russia and launched on Russian rockets.

When a company named Payload Systems, founded by former astronaut Byron Lichtenberg, wanted to fly a protein crystal growth project to the Russians' *Mir* space station, the OSC championed the company's cause and prevailed against NASA's opposition.

In December 1989 Manber visited the Soviet Union to be on hand for the memorable event, the launch of the first U.S. commercial science payload carried aboard a Soviet booster. "The Russians were delighted when the Payload Systems project got approval," Manber recalled. "They saw it as a sign—which it was not—that Reagan's people were willing to work with the Russians on a free market in space."

Manber, who had left the OSC by this time, was approached by Energia, the Russian manufacturer of spacecraft and the *Mir* space station, to help it become a company and develop its commercial potential. NPO Energia held a unique status in the Soviet space program. In many respects it operated like NASA: it designed, developed, built, and operated its own programs. It built and operated the *Mir* space station. Increasingly, it was also being expected to fund its programs.

Manber would accept Energia's offer in 1992 with the creation of Energia Ltd., an office of Energia that facilitated U.S.-Russian cooperation in space and the development of contracts with American space companies.

If the Soviets were not getting their share of the commercial satellite launch business, they had cornered the market in flying private citizens into space. In 1990, when the Houston Space Society wanted to change the way people thought about space, the Russians already had two deals brewing to send private citizens to the *Mir* space station. The Tokyo Broadcasting System (TBS) had paid a reported $12 million to put one of its reporters, Toyohiro Akiyama, on a flight to *Mir*, scheduled for later that year. Meanwhile, in the United Kingdom, a private group was trying to raise the necessary funds to send the first Briton into space.

It wasn't surprising, therefore, that in their brainstorming for a path to space, the Houston Space Society members immediately leapfrogged over the U.S. space agency to reach the Russians. The Russians offered a way to get to space, with just one sticking point: how could the Houston Space Society come up with the $12 million price tag? They first thought of a lottery, but then society member David Mayer suggested instead a 900-number sweepstakes. He pointed out that just the previous year, Music Television (MTV) had conducted a 900-number sweepstakes to sell the boyhood home of rock star Bon Jovi. Each of the eight hundred thousand callers to that phone number was charged 99 cents to be entered into a lottery to win the house. Just imagine how many would call if the prize was a trip to space!

Although the Houston Space Society eventually decided not to pursue the idea of sending someone into space, Mayer and Davidson continued to discuss the idea until they became convinced it would work. Perhaps two guys from Houston really could put a person into space. And why stop with one trip? Why not create a travel agency to send more and more people into space, maybe eventually to the moon? They were on fire with the possibilities. Over the next few weeks they researched how sweepstakes worked, contacted a sweepstakes administration company, got bids for advertising the contest, and wrote up a business plan. Now they had to arrange the launch.

By happy coincidence, the person to speak to in 1990 if you wanted to arrange anything with the Russians lived in Houston. Attorney Art Dula had been involved in the development of commercial space since 1980, when he helped Houston real estate mogul David Hannah Jr. establish Space Services, Inc. of America (SSI). Hoping to be a private satellite launch company, SSI hired former Mercury astronaut Deke Slayton as its president, ran a gauntlet of eighteen federal agencies to get permission, and in 1982 launched the first privately financed rocket into space.

In the wake of the *Challenger* disaster, Dula traveled to Russia and encountered a new mood of openness, a new willingness to adopt Western-style attitudes toward commercializing its space program. Dula and his team became the first Westerners ever to visit the super-secret launch complex, the Baikonur Cosmodrome. Dula actually laid his hand on a Proton rocket sitting on the launchpad. In a complete reversal from their traditional cold war secrecy, the Russians were now prepared to share technical information about their satellites, rockets, and space-based cameras.

Back in Houston, Dula created the Space Commerce Corporation and began marketing the Proton rocket as a launch vehicle. The job would have been a lot easier had it not been for restrictions on sending U.S. technology to the Soviet Union. "They're trying to open up," Dula reported after his visit to Russia. "They want to be capitalists and we're trying to help them." The issue boiled down to whether the United States was going to allow them to be capitalists.

Fortunately, restrictions on exporting U.S. technology to the Soviet Union did not apply to human beings. Mayer and Davidson met with Dula, and he agreed to pursue the matter on his next trip to Moscow. As a backup plan, the pair also met with Eagle Engineering. Composed largely of ex-NASA engineers, Eagle Engineering had developed SSI's rocket, the Conestoga, as a private satellite launch vehicle. If you had the money, they could get you into space.

According to Davidson, he and Mayer met with Eagle to discuss the possibility of putting together an Atlas rocket to launch a Mercury-style capsule for an orbital flight. This would be their backup plan if the Russian arrangements didn't pan out.

Eagle's estimate for such a "man rated" safe mission was $1 billion. Mayer caught his breath when they mentioned that figure and countered with, "What could I get for a hundred million?" The way Davidson recalls that meeting, the Eagle people were stunned. When they regained their composure, they suggested that a lunar sample retrieval mission might be put together within that budget.

Fortunately for the venture, Dula returned from Moscow in November with a signed contract, allowing preparations to shift into high gear. Davidson and Mayer established Space Travel Services Corporation (STSC). Within weeks the pair had raised seed money, hired a publicity firm, and made arrangements with a phone company. The only glitch in preparation occurred when the company that was to operate the sweepstakes unexpectedly pulled out. According to Davidson, they would later learn that the company had been told by someone of authority in Washington that this was something that would not be allowed to happen.

Undaunted, STSC hired an accounting firm to handle arrangements and plunged ahead. On 17 December 1990 STSC held a news conference to announce the sweepstakes, "The Ultimate Adventure." Anyone could buy a chance on going into space by calling a 900 telephone number at a cost of $2.99. "He or she need not be a test pilot or a scientist or a government official," claimed a company statement. "The butcher, the baker, the candlestick maker are all eligible. It could be your neighbor. It could be you."

The announcement drew national attention. In short order, call volume shut down several telephone exchanges at the Dallas offices of the phone company taking the calls.

The timing for the Ultimate Adventure could not have been better. Exactly one week before the press conference, Japanese journalist Toyohira Akiyama returned from his eight-day visit to *Mir*, thus demonstrating that the Soviets were indeed taking paying customers into space. Akiyama's news company, TBS, had paid for his ticket to fly.

The forty-eight-year-old foreign news editor for TBS had been found medically worthy and had successfully completed eighteen months of cosmonaut training, despite his recently kicked habit of smoking forty cigarettes a day. If anything, his flight proved that even those with considerably less than the "right stuff" could travel in space.

Allowing commercial customers into space short-circuited the more traditional selection process. Just as NASA resisted STSC's astronaut sweepstakes, Akiyama's flight led to protests.

The Japanese Aerospace Exploration Agency (JAXA), which had close ties with NASA, had hoped that one of its astronauts then in training for a shuttle flight would be the first of that nationality to fly in space. JAXA made clear its anger at having this honor snatched away by a journalist who bought his way into space on a Soviet rocket.

Soviet journalists were also incensed that a foreign journalist would have the honor of being the first of their profession in space. In 1965 the Russians had actually recruited a group of three cosmonaut journalists. When the Voskhod mission they were to fly on was canceled, they missed their opportunity. It was a matter of national prestige, the journalists now claimed. The first man, woman, doctor, and engineer in space had been Soviet nationals. Why not the first journalist as well? They accused the Russian contract agency for space affairs, Glavkosmos, of being "mercenaries."

In earlier days Soviet space officials might have taken offense at such a charge; now it was downright complimentary. It simply made sense to utilize all the assets at their disposal. The simple equation went like this. The Soyuz spacecraft contained three seats, two for crew and one spare. In the past, that extra seat had often been given away to someone from a Soviet bloc or nonaligned country as a gesture of goodwill. But by the late 1980s, with a crippled Soviet economy, it occurred to officials that the extra seat could be a source of much-needed hard currency to replace vanished government funding.

So it was with a sense of pride rather than offense that Glavkosmos announced not only that would it be taking paying passengers into space but it would also be selling advertising on its space station *Mir*, and that astronauts, like tennis players, would soon be wearing sponsors' patches on their space suits.

In fact, this is precisely what the deal with TBS entailed. For its $12 million investment, TBS and its sponsors hoped to reap an avalanche of publicity. They stretched out programming for the full eighteen months of Akiyama's cosmonaut training by filing reports of his progress and his struggle to give up cigarettes. When the *Soyuz TM-11* rocket lifted off from Baikonur, it was adorned with images of the rising sun and the corporate logos of a Japanese pharmaceutical company and a manufacturer of sanitary napkins. Other advertiser logos decorated the launchpad and the cosmonauts' T-shirts.

As the first journalist in space, Akiyama conducted a nightly broadcast to report on his adventure. These broadcasts got off to a poor start when he used his first live report to discuss the nastier effects of his space sickness, which felt like a bad hangover.

Among his other official duties on the flight were filming several commercials and videotaping his activities. The tapes included a segment on how toys behaved in zero gravity and a segment on a scientific experiment, called Frogs in Space, to demonstrate how Japanese tree frogs behaved in zero gravity. The experiment had been selected for its visual appeal and because the frogs were popular with children.

Akiyama's broadcasts from space drew high ratings and put a human face on the role of the space traveler. His brain felt like it was "floating around in my head," he reported. He had neglected to pack enough underwear. Here was an ordinary person, getting sick, craving cigarettes and his favorite foods, and excitedly reporting that he could see Mount Fuji from space. Clearly, space had become a very exclusive yet viable tourist destination.

As with the earlier group of passenger scientists, such as Charlie Walker, Akiyama's flight grew out of the trend to commercialize space. As such it bears similarities to Walker's flight or the motivation behind the Teacher in Space program: to derive practical benefit from the billions of dollars invested in space and to generate publicity that increased public support for the massive budget that space required.

However, in the case of Akiyama, for the first time a space agency sent a passenger into space as a way to help pay the bills. In the 1980s the Soviets used their Intercosmos program as a diplomatic perk for their Eastern European allies. But it eventually evolved into a paying proposition. Any country could have its own astronaut if it covered the cost. Akiyama's flight developed naturally from that premise. If countries could buy their way into space, so too could commercial entities or even individuals.

Akiyama's flight also served as a wonderful advertisement for Davidson and Mayer's Ultimate Adventure, proving that such flights could be arranged with the Russians and that an ordinary individual could make a trip into space.

No sooner had media coverage begun to generate considerable interest in the sweepstakes than legal problems threatened to scuttle the project. The Soviet news agency TASS reported that the sweepstakes was a hoax because no such agreement existed to send the winner into space. On that same day, Harris County and Texas state district attorneys announced they would investigate the sweepstakes.

Given the fragmented nature of the Soviet space program in those years, this confusion wasn't surprising. The Russians had no single space authority but rather a collection of loosely connected authorities. Dula had received a letter from Energia, the manufacturer of spacecraft and space stations, which managed *Mir.* TASS had spoken with the Soviet space agency, Roskosmos, nominally in charge of the country's space program. The charge that it was a hoax appeared in a 19 December *New York Times* article. Another article the following day cleared up the matter by quoting the deputy director of the foreign trade subsidiary of NPO Energia, Vladimir Nikitsky, who claimed the deal was legitimate.

Meanwhile, both Jim Davidson and David Mayer ended up in the Harris County jail on 7 February, charged with operating an illegal lottery. Over the next several months, the pair battled the state and county attorneys general over charges they had violated Texas anti-gambling laws. Some ten thousand calls would eventually come into the 900 number, and deals for television and advertising rights were in the works. But legal fees also mounted, and the first installment of $1.5 million was due to Dula's Space Commerce Corporation by 1 April 1991. The end came amid a swirl of crim-inal charges and civil injunctions. By May 1991 STSC, the first company created to send a private person to space, was out of business.

The irony of the failure of STSC to arrange private financing for a spaceflight was driven home that same month when Helen Sharman, a private British citizen, launched on a Soviet flight to visit *Mir*.

The impetus for her flight emerged in the heady year of 1989. Only a few months after TBS signed with the Soviets to fly its reporter, Moscow Narodny Bank, a Soviet-owned bank in London, launched a project to put together a privately financed flight for a British citizen. The success of the Akiyama deal suggested that spaceflights might be marketed much like other high-profile entertainment or sporting events. However, unlike the flight of the Japanese journalist, the British project planned to recruit a private citizen, sell corporate sponsorships, and fly British science experiments.

The call for a private astronaut in Britain touched on a matter of national pride the way it never could have in the United States. Through its political goodwill program, Intercosmos, the Soviets had already flown a Mongolian, an Afghan, a Cuban, and a Syrian. However, in 1989, no Briton had traveled in space.

Narodny Bank provided the seed money to create a British company named Antequera to oversee British investment and manage the project. The cost of a visit to *Mir* was put at $12 million. The cost of developing British scientific experiments and integrating them into the space station would bring the overall cost to $16 million. The world's largest advertising firm, Saatchi & Saatchi, was hired to sell corporate sponsorships and broadcast rights, and to recruit a British astronaut. Given the name Project Juno, the enterprise got under way in June 1989, when advertisements appeared in print and on radio.

The day it all began, Helen Sharman, a chemist working for the Mars Confectionery Company, was stuck in a traffic jam on her way home from work. She flipped through the radio dial, looking for music, and instead heard an advertisement asking for volunteers to go into space: "Astronaut wanted; no experience necessary," it proclaimed.

"I had no background in space research," Sharman would later recall. "I had never harbored ambitions to go into space. It had never occurred to me that I could or would ever wish to go into space, let alone be able to."

However, she did meet the necessary qualifications: She was a British citizen, in the age range of twenty-one to forty, with formal scientific training, a proven ability to learn a foreign language, and medically fit. Sharman had been born twenty-six years earlier in Sheffield and held a degree from Sheffield University. She was studying part-time for a PhD at Birbeck College, London, when she took a position as a research chemist with Mars Confectionary.

Along with 13,000 of her fellow citizens, Sharman called the phone number provided and, after a few screening questions, was sent an application. A review of applications quickly reduced the number of candidates to 150.

Unlike the U.S. STSC, Project Juno was not planned as a sweepstakes. Sponsorships, advertising, and various media deals would finance the project. The astronaut would be chosen by a vigorous selection process. By August, increasingly exhaustive medical examinations had further thinned the ranks of would-be cosmonauts to thirty-two. The candidates were given a taste of things to come that month when they were all brought together for the first time.

Various speakers told the thirty-two about the project, the selection process, and the likely science experiments that would also fly. Publicity was key to the success of Project Juno, someone from Antequera explained, bluntly stating they would "own the butt" of the candidate finally selected. Then the candidates were ushered into a room and for the first time thrown to the waiting press. Because of her connection to the Mars Confectionery Company, Sharman was quickly labeled the "Girl from Mars."

Media interest remained high as the selection process stretched over many months. When the process had winnowed their numbers to four, the remaining candidates traveled to Moscow for more medical examinations. On their return from Moscow, millions of British viewers tuned in to a TV special that would announce Britain's astronauts, the lucky two who would actually train for the mission.

Sharman remembered it as an "excruciating," beauty pageant–type production, a "Monty Python sketch." Dressed in poorly fitting "flying suits," the four candidates sat before a studio audience. Then the broadcast went live to Moscow, where one of Antequera's directors announced that Army

Air Corps Major Tim Mace and Helen Sharman were the two finalists who would train in Star City, while the other two would remain in Britain as the backup crew.

For many agonizing minutes they stood on stage smiling and waving while the audience applauded, and 5.5 million viewers drank in the scene. Then it was off to another press conference and a champagne reception to sign hundreds of autographs and greet directors from Antequera, Glavkosmos, and even Mars Confectionery.

The warm glow of British-Soviet cooperation came against the backdrop of the dramatic changes unfolding in the Soviet Union and throughout the communist world. Sharman and Mace arrived in Star City to begin cosmonaut training on 30 November 1989, the same month the Berlin Wall fell. If this symbolic reconnecting of the West and East gave heart to the team at Antequera, that new vision was not shared by the corporations they solicited for project support.

All of the foreign citizens who trained at Star City, be they NASA astronauts, Intercosmos participants, or private individuals, faced difficult periods of adjustment. That was certainly the case for Sharman and Mace. The facilities were austere. Not knowing the language and dealing with strict training rules increased their feelings of isolation.

Their training began with learning the Russian language. Along with additional medical evaluations and physical conditioning, this occupied the first few months. Even with the rudiments of Russian mastered, it was still a challenge to follow classroom work in the physics of space flight. That segued into mastering such practical skills as parachuting, moving in a weightless environment, and emergency drills. In the later stages, work shifted to crew training. This involved working in simulators, practicing launch and reentry procedures, learning the layout of the Soyuz and *Mir* spacecraft, and becoming familiar with the science experiments that would fly with them.

Three months into their training, an Antequera representative paid them a visit with some sobering news. The project's advertising agency, Saatchi & Saatchi, had approached fifty companies about sponsorships, but so far had signed only three. If they didn't meet with more success, and quickly, the entire project might be canceled.

This uncertainty hung over their heads for most of 1990. News stories reported claims by Saatchi & Saatchi that they were having trouble identifying exactly what would and what would not be possible with the Russians. If they sold sponsorship to a pocket calculator company, would the Russians allow a pocket calculator onboard? If a communications company wanted a sponsorship, would its equipment be compatible with *Mir*'s signals? There was never a complete resolution of the benefits of sponsorship.

As 1990 progressed, it became clear that Saatchi & Saatchi would not be able to cover the costs of the mission with commercial sponsorships. Moscow Norodny Bank eventually took over the project. Although the bank managed to sign up three additional sponsors, including Interflora, a flower delivery service that wanted a flower order placed from space, it raised only $1.7 million of the projected $12 million cost of the flight.

In the end, the Soviet government chose to save face by covering the costs of the project. The sponsorships would be honored, but corporate logos would not adorn the rocket or the launch site. Also, to cut costs, the original British scientific experiments planned for the mission were scrapped and replaced by less expensive ones designed by teachers and schoolchildren.

The fact that the Project Juno astronaut would no longer be conducting British science experiments in space caused a shift in the public's perception of the project. Antequera's sponsors pulled out, since they could no longer be offered anything for their money.

None of this had any impact on the mission, however. By December 1990 Sharman and Tim Mace had begun the crew training preparation for the flight. Increasing amounts of time in simulators and practicing the many flight procedures made those routines second nature to them. By the time they began the scientific phase of training, learning how to handle the Russian experiments that would be on the flight, another anxiety began to weigh on their minds. As Sharman recalled in her autobiography, "Both Tim and I knew, however, that the time was fast approaching when one of us would be selected for the prime crew, the other for the backup."

In mid-February 1991, Project Juno officials visited Star City. First they interviewed the Russian trainers and then the two candidates. How did they feel about the training? they wanted to know of Mace and Sharman. How did they feel about the mission? Who did they think should be the selected candidate? Why?

Later that month the pair flew back to Britain for the final selection, where Sharman learned that she would be on the prime crew and Mace would be her backup.

Sharman's parents and the British ambassador to Russia attended the launch on 18 May 1991. Sitting atop the Soyuz rocket, with her cosmonaut crewmates Anatoli Artsebarski and Sergei Krikalev, Helen Sharman imagined how the launch would look and sound to her parents. She had been on hand to watch the launch of Toyohiro Akiyama and knew herself how the roaring rocket looked and sounded as it climbed into the sky, and the heightened anticipation of those watching from the outside.

In her autobiographical book, *Seize the Moment*, Sharman reported feeling calm as she awaited liftoff, even as her mind raced over the dark portent of events and conversations of the past few days. Her thoughts also turned briefly to another female civilian astronaut, Christa McAuliffe, America's teacher in space, who died in the *Challenger* disaster. Over the past few months, journalists had often asked for her thoughts about the astronauts who had died on the *Challenger*, apparently to draw a comment about the danger of her mission.

In the final minutes, Sharman cycled through all of the similarities between their circumstances and the unlikely prospect that history would repeat itself in another disaster. In the end, the launch went flawlessly.

Back home, London's *Observer* newspaper ran the launch story under the headline, "Woman from Mars Is First Briton in Space." Despite the media attention given to Project Juno, however, only about half of Britain's newspapers carried the story on their front page. Some of the luster had worn off the project. British companies were not paying for Sharman's flight, and no British science was being conducted. Britain had become the twenty-second nation to have one of its citizens in space.

In orbit, Sharman conducted Soviet-designed experiments, took photographs of Earth for a remote sensing project looking at pollution in major estuaries, and contacted nine British schools by ham radio. She was photographed using a handheld device manufactured by Poqet Computers, one of the sponsors. For the sponsor Teleflora, she radioed to Earth an order for flowers, which she sent to her mother.

One of the most harrowing experiences of the mission occurred when the crew's meal was interrupted by a series of insistent beeps, signaling a loss of power in the station's batteries. Power loss happened routinely when *Mir* passed through Earth's shadow, robbing its solar panels of sunlight. Still, it was unnerving as the noisy ventilating fans across the station began to stop, bringing an eerie silence. Then the lights went out, leaving only a single emergency fluorescent tube glowing.

Even though Sharman had been prepared for this and knew exactly what was happening, she would later report that "it felt, unmistakably, as if the station had started to die and it gave me a cold, sinking feeling." Waiting for the station to emerge once again into sunlight and regain power, she had the sense of drifting helplessly in what she characterized as a "lifeless hulk."

Mir was showing its age. It had already survived beyond its intended lifetime of five years. It would continue to support missions for ten more years and ultimately find itself at the center of a heated controversy over the commercialization of space.

The visits to *Mir* by Akiyama and Sharman created the template for how private individuals could travel into space. However, Sharman and the whole Project Juno experience would be far more representative of private efforts to take other visitors to *Mir*. There was no shortage of people eager to go, or of schemes to pay their fare. But the Russians had become increasingly wary of grand schemes to raise flight money. They still wanted to transport paying customers into space, but they wanted more credible payment plans, just the sort of payment plans offered by NASA.

In September 1993 an agreement between the United States and Russia laid the framework for cooperation on a new space station, which would eventually become known as the International Space Station (ISS). From 1994 to 1998, cosmonaut flights on the shuttle and astronaut visits to *Mir* began to develop the collaborative relationship required for joint construction of the ISS.

These years of U.S.-Russian cooperation aboard *Mir* were alternately labeled the Shuttle-*Mir* program or "Phase 1." The latter term attempted to establish the program as a preliminary step in the construction of the ISS (Phase 2) and make the program more palatable to the U.S. Congress, which was essentially being asked to fund Russian space activities following the collapse of the Soviet Union. This clever way of packaging the project allowed for the continued operation of *Mir* and provided desperately needed funding for the Russian space program.

These years were remarkably active. Eleven shuttle missions flew to *Mir*, seven astronauts made long-duration visits, the station acquired new modules, and the two nations learned how to work together to assemble structures in space. However, the condition of the station during these years reflected Russia's dire economic situation. It became infamous for its deteriorating condition, harrowing accidents, and inadequate budget. The Russian Space Agency (RSA) continued to offer "guest cosmonaut" flights to foreign governments for substantial fees. The French and German space agencies each paid $14.7 million to send up one of their astronauts.

By 1997, however, the RSA and Energia had become even more inventive in marketing their services. Western tourists began to appear at the Gagarin Cosmonaut Training Camp in Star City for a $1,000 one-week "space training course." *Mir* commander Vasily Tsbilyev starred in an onboard commercial for Israel's biggest food manufacturer, TNAVA, by swallowing a floating globule of milk, squeezed from a Hebrew-lettered carton. Pepsi paid to have *Mir* cosmonauts deploy in space a giant replica of its new blue can.

It also appeared as though *Mir* was about to return to the business of hosting private citizens. The TV news network CNN entered into preliminary talks with RSA about sending its science correspondent John Holliman to *Mir* for a series of special broadcasts. News reports also began to appear about British businessman Peter Llewellyn agreeing to pay the Russians $100 million for the opportunity to fly to *Mir*. It always came down to money, however, and despite the best intentions of all of these ventures, there was never enough. CNN could not find sponsors, and Llewellyn promised a lot more than he could deliver.

As soon as the final astronaut, Andrew Thomas, returned from *Mir* in June 1998, NASA began pressuring Russia to deorbit *Mir* as soon as possible and concentrate its limited resources on the ISS. There was an undeniable logic here. If it was continuously occupied, *Mir* cost $200 to $250 million a year to operate and about half that amount if it was empty. But allowing the venerable space station to burn up in the atmosphere did not go down well with the Russians. Aside from being a very valuable resource, the station had become a symbol of past glories in these hard economic times.

Enter Russian movie mogul Yuri Kara. In 1998 he launched a film project designed to put an actor aboard *Mir* for a space movie that would premier in April 2001, the fortieth anniversary of Yuri Gagarin's historic space flight. "This movie is very important not only for the art but for the state," claimed Kara. "The first man in space was Russian, and it would be really symbolic if the first actor in space was Russian as well."

Titled *The Final Voyage*, the plot revolved around a renegade *Mir* cosmonaut who refuses to return to Earth. Ground control sends a woman into space to persuade him to return. The project plans called for recruiting cosmonauts aboard *Mir* to act as cameramen to film the actor while director Kara monitored the action from mission control.

Finding a suitably qualified Russian actor proved to be the first of many challenges. Four men and three women competed for the honor. Some would not commit to the rigorous and lengthy cosmonaut training. Others failed the extensive medical screening. "What could you expect from actors whose lives consist of festivals, acting, parties and drinking?" Kara lamented.

Russian actor Vladimir Steklov eventually won the role and completed four month of cosmonaut training in 1999 with cosmonauts Sergei Zaletin and Alexander Kaleri, who were slated for a March 2000 mission to *Mir*. The project gathered momentum when British producer John Daly was brought in and American actors Gary Oldman and Sean Penn signed up for the on-Earth scenes of the film.

"I don't want to spoil the plot," actor Steklov confided to a reporter, "but I play a cosmonaut who refuses to leave the *Mir* and instead makes an address to humanity from space. My lines are: 'We need to stop and realize who we are, where we come from, why we exist, and what we are doing to our common home — the Earth.'"

All of this played out in late 1999 against a backdrop of uncertainty regarding *Mir*'s fate. The simple truth was that the Russian government could no longer afford the luxury of *Mir*. A final mission flew to *Mir* to install a computer that would allow ground control to deorbit the station. And yet the government held out hope that outside funding could still be found to save the station. Russian president Boris Yeltsin approved keeping *Mir* in orbit if the government did not have to pay for it.

Various politicians and space officials appeared in the Russian media seeking to build support for continuing *Mir*. In a desperate move indicative of the affection felt for *Mir*, a group of former cosmonauts led by Vitaly Sevastyanov opened a bank account and urged the economically hard-pressed Russian population to contribute to the fund to keep *Mir* flying.

As he explained, "It is a space station made in the Soviet Union that is continuing to work after the Soviet Union is gone. It is a symbol of what we could once do."

The RSA handed out calendars for the year 2000 with a picture of *Mir* captioned, "Will we still fly?" As RSA spokesman Sergei Gorbunov explained, "Everyone wants this station to fly. It is like a badge of honor for Russia. Unfortunately it's a very expensive one."

Financial backing for *The Final Voyage* (alternatively titled *The Last Survivor*) proved as difficult to find as funding for *Mir*. In fact, funding *Mir* and funding the movie proceeded along parallel paths in late 1999 and early 2000, each a reflection of the other, each hoping for a miracle benefactor. The movie had a production budget of $200 million, with $40 million budgeted for the in-space portion. The film's producers made an initial payment to the RSA and Energia, and Steklov completed training. But despite frantic last-minute transatlantic visits and phone calls, sponsorship funding did not materialize. It was an all-too-familiar story.

Kara asked Energia to allow Steklov to undertake the mission anyway, arguing that once he had acquired the in-orbit film, it would be easier to find the additional funding and pay the balance due. But Russian space officials would have none of that. It sounded too much like "Project Juno" to suit them. By now they had heard every imaginable promise and scheme for paying for passage to *Mir*. *Mir* was scheduled to deorbit. They didn't need promises, they needed cash. One month prior to the March 2000 launch that would have taken Steklov to *Mir*, he was removed from the crew manifest.

What unfolded next in the long, convoluted history of *Mir* might well have come from the plot of an action movie. *Mir* had been unoccupied since August 1999, like a condemned criminal awaiting the inevitable death sentence. The much publicized Shuttle-*Mir* program had concluded. Based on a few harrowing accidents and a spartan Russian space budget, *Mir* had acquired the reputation as an unreliable piece of aging space hardware that had outlived its usefulness. *Mir*'s fourteen glory years as the premier space station, the only space station, were at an end. The Russians had signed on as partners with NASA on the ISS.

Ownership of *Mir* remained with the Russian government, but the Energia Corporation had become the designated operator of the station and its last chance to find outside funding. *Mir*'s eleventh-hour reprieve came in the form of MirCorp, a consortium of Western multimillionaires and space visionaries who wanted to acquire *Mir* for a variety of commercial ventures.

Two of the leading figures in MirCorp were telecommunications entrepreneur Walt Anderson and director of the Space Frontier Foundation Rick Tumlinson. The two shared the deep conviction that NASA had the whole space thing wrong.

Born in 1953, Anderson grew up under the influence of the grand accomplishments of the space age — the first satellites, manned flights, moon landings — and those moments inspired his obsession with space. He was a man who acted on his obsessions. Flush with telecom industry wealth, Anderson had long championed private enterprise in space. He had helped to finance the International Space University in 1989, established the Space Frontier Foundation in 1994, underwritten the creation of the Foundation for the International Nongovernmental Development of Space in the mid-1990s, and invested more than $40 million toward the development of Roton, a manned, reusable spaceship.

Tumlinson, on the other hand, had deep L5 Society roots from the early 1980s. He had worked with Gerard O'Neill in Princeton and Freeman Dyson at the Space Studies Institute. He sometimes referred to himself as one of "Gerry's kids" or "Apollo's children," to indicate his conviction that the goal of our space program should be to lead to space settlement. The role of NASA should be exploration. Period. Get us into orbit, land us on the moon, and then get out of the way and let private enterprise take over. Let private individuals go into space. For Anderson and Tumlinson, *Mir* presented a golden opportunity to advance their agendas.

In the spring of 1999, with news stories appearing regularly about the demise of *Mir*, Anderson and Tumlinson met at a Los Angeles restaurant to consider their options. *Mir* had received so much bad press recently that everyone thought it was a piece of junk. These two thought otherwise. "Sure, the wiring was bad and it had some rust and leaks," Tumlinson conceded. "But if you had one of the most expensive buildings on the planet and the wiring and plumbing were worn out, would you tear it down?"

Mir still had years of useful life, they argued. Companies would pay to use it. People would pay to visit it. Advertisers would pay for a piece of space. One business analyst had put the value of *Mir* at $1.2 billion. In a pure business sense, *Mir* was an undervalued asset. What if they bought *Mir* and developed all of its business potential? A private company operating its own space station? The sheer audacity of it was irresistible.

The most urgent problem confronting them was the slow decay of *Mir*'s orbit. If something was not done soon, there would be no *Mir*. Their solution, sketched out on the restaurant tablecloth, would be to dangle forty-four miles of copper wire below the station. This innovative technology, known as an electromagnetic tether, would pump electricity down the wire and use reverse polarity from Earth to push *Mir* to a higher orbit.

Step two of the plan would be to send a crew to *Mir* to check it out and shut it down, essentially parking it safely in this higher orbit for a year or two until they could decide what needed to be done.

Under the looming deadline of a *Mir* deorbit, the pair enlisted Jeff Manber to help them buy *Mir*. They wouldn't be able to buy it, Manber explained, since it was Russian government property, but they might be able to lease it. In a hastily arranged Moscow meeting with Energia, the company that managed *Mir*, the group found the Russians cool to the idea. The Russians wanted to save *Mir*, but, as Manber saw it, they were just tired. They had suffered through their share of ill-conceived schemes to rescue the station and been duped by their share of schemers. They didn't have the money to keep *Mir* up, and NASA was exerting considerable pressure on them to deorbit the station. Manber and Tumlinson managed to come away with a nonbinding agreement to work on a plan to save *Mir*.

That was enough to energize Anderson's team. Washington attorney Jim Dunstan well remembers the next step in the process. Dunstan belongs to a small fraternity of attorneys who specialize in space law. He had taken a course in space law at Georgetown, founded the Georgetown Space Law group in 1982, and has been at it ever since. Prior to a return trip to Moscow, Anderson asked Dunstan to write up a lease agreement for a space station.

A private company leasing a functioning space station to turn it into a paying proposition — it was unique, revolutionary. How did you draw up a legal document for that? But Anderson kept saying, "It's a simple business deal. Keep it simple." That thought resonated with Dunstan as he put the pieces together. He repeated to himself the basic facts. "Energia had control of *Mir*, five modules up there, sitting in orbit. Then it suddenly dawned on me, 'five rooms with beautiful view.' I literally went over to my shelf and took down one of the legal form books and pulled out a standard apartment lease. 'We hereby agree to lease such and such a property, located at this orbital inclination.' The whole thing was very short."

Anderson went along for the next trip to Moscow and hit it off right away with Energia head Yuri Semyonov. Both were men of action. They hated rules and bureaucrats and government oversight. They were pragmatists who valued results. Results were precisely what Semyonov needed from MirCorp as he laid out the undeniable facts of that moment in the life of the *Mir* space station. "When I leave this room," he explained, "I have to give a command to mission control. That command is either going to tell my engineers to press a button that fires a thruster to push *Mir* down and begin its deorbit, or another thruster that pushes it up, to keep it in orbit. Now, show me that you're real."

Anderson's team experienced a collective lump-in-the-throat moment. On their flight to Moscow they had played the board game Risk, the game of world conquest. In a symbolic way, it represented their thinking on the magnitude of what they were trying to accomplish. However, sitting in a Moscow restaurant before meeting the Russians, Anderson had asked them all to estimate the likelihood that they could actually pull off a deal to take control of *Mir*. Most guessed their chances on the down side of 50 percent. Team member Gus Gardellini pegged the chances somewhere below zero. Anderson himself proved the most optimistic, coming in at 60 percent. Now, the moment of truth had arrived. They had been trying to convince the Russians that they could come up with the financing necessary to sustain *Mir*. Now they had to deliver.

"It was an amazing moment where he [Anderson] had to decide," Tumlinson recalled. "Could we go for it and would he be willing to put up several tens of millions of dollars to make it go."

Anderson did not speak Russian, but he understood the language of business. He understood that a dramatic gesture would demonstrate the kind of individual he was, the sort that could cut through the labyrinth of red tape, thumb his nose at NASA, and still overcome the mountain of obstacles that stood in their way. In short, that he was the kind of individual in whom the Russians could place their trust.

Anderson had already arranged with his bank that when he phoned them and mentioned a key phrase, he could authorize a transfer of funds.

At a meeting with Energia later that same day, Anderson whipped out his cell phone, called his bank, spoke the words "*Mir* for peace," and with that wired $7 million to Energia's bank account. All this before they had a signed agreement.

For Tumlinson that was the moment when the reality of the project hit him. "I remember several of us looking at each other and going 'Oh, my god.' This is really going to happen."

That gesture won the Russians' trust, and the deal came together. Now they needed a legal document to define their agreement. Anderson slid across the table the very thin contract drawn up by Dunstan, which had been translated into Russian.

"What's this?" Yuri Semyonov asked. Anderson answered softly, "An apartment lease." Semyonov looked at him for a moment as that thought sank in, then burst out laughing. Additional addendums and protocols would have to be added, but essentially that landlord-tenant agreement defined their relationship. Energia would hold 51 percent of *Mir*, and as landlord would be responsible for "building maintenance," all of the mechanical and political matters involved with maintaining the residence. MirCorp, as the tenant, got rights to *Mir* for the rest of its life, the use of two or three manned Soyuz rockets per year and two or three unmanned Progress cargo rockets, plus exclusive control of visitors.

MirCorp had taken possession of a piece of orbital real estate. It was a Gerard O'Neill moment of the highest magnitude. Now they only had to come up with the rent for that "five rooms with beautiful view"—$200 million per year.

"We went from a cold start to leasing a space station in the span of sixty days," explained Dunstan. But now the real work began. The job before them was herculean and made more so when they learned that the tether they planned to use to boost *Mir*'s orbit fell under ITAR regulations, which prevented the export of certain technologies. If they could not use the tether to boost *Mir*, they would require an additional rocket launch. It was just one of the many hurdles awaiting them.

On that memorable trip to Moscow, Anderson asked Jeff Manber to head up the company, MirCorp, that would manage the station. Manber turned him down for two reasons. One, he didn't think it would work, and two, "I knew that whoever was in charge of MirCorp would really feel the wrath of NASA, and I just wasn't ready to do it knowing we would fail."

Never understanding the meaning of no, Anderson persisted. He needed someone up to the challenge and with whom Energia would be comfortable. As Manber recalled it, his final response to Anderson went something like this: "Walt, I don't believe we will have commercial success. I believe that NASA politically will kill this project. I don't wish for you to burn through $40, $50, $60 million, unless you can tell me that one sign of success for MirCorp will be that if at the end of the day, when this project ends, I'm able to get enough attention and enough publicity that people think differently about private space and commercial markets than when we started. Will you consider that money well spent? He looked down and quietly said, 'Yes I would.' And I said, 'All right, I'll take the job.'"

The world media turned out for the January 2000 press conference in London to announce the formation of MirCorp. Energia director Yuri Semyonov lent gravity to the event as Manber reviewed for the press this totally unprecedented arrangement of a private corporation leasing a space station. Although the reporters could appreciate the sheer novelty of the venture, they had trouble grasping the paradigm shift involved with private space.

Jeffery Manber vividly remembers the most telling moment of the press conference. "When I got to the part of our business model that said we were going to seek out and sign up rich people to fly into space, the reporters burst into laughter." In 2000 the idea that rich people would pay enormous sums of money to risk their lives in an aging Russian space vehicle seemed too ridiculous to be credible.

From that point it was a race against considerable odds. Tumlinson compared it to being thrown into a pool of sharks while somebody shot at you with an Uzi submachine gun. The race was on to find customers and other investors willing to cover the enormous costs of running a space station. And this had to be done against the backdrop of the relentless decay of *Mir*'s orbit and the efforts of both the U.S. and Russian governments to bring down *Mir*.

In the United States, NASA administrator Dan Goldin came close to meltdown during congressional testimony. "I'm highly frustrated by the fact the Russians signed up to a program that put such incredible emphasis on *Mir*," Goldin complained to a House subcommittee. "They always seem to have a little extra money around for *Mir* but not the International Space Station."

It was as though the ISS mega-project would now have to make room in orbit for the upstart commercial project that threatened to skim off resources from the Russians. That dose of free-market competition did not sit well with NASA. The Russian Space Agency, eager to preserve its good relations with NASA, also wanted to forget about *Mir* and concentrate on ISS. Each agency would work behind the scenes against the success of MirCorp.

Set in that context of political pressure, the MirCorp team had to generate revenue very quickly. In fact, they had to create the entire company from scratch: hire staff, arrange financing, develop a business plan and business relationships, create an image and a product line, and so on. Aside from space tourism, MirCorp planned to sell access to the station for corporations and countries, sell advertising, sell the intellectual property rights to *Mir*'s many patents, and eventually get into in-orbit satellite assembly and repair.

With an initial investment of $20 million from Anderson and additional funds from fellow telecom tycoon Chirinjeev Kathuria, MirCorp scored two quick successes. In February 2000 a Progress rocket boosted *Mir* to a higher orbit, extending its lifetime.

In April MirCorp made history with the first commercial manned space flight. Completely financed by MirCorp, a Soyuz spacecraft carried cosmonauts Sergei Zaletin and Alexander Kaleri to *Mir* to bring it out of its hibernation and prepare it for human occupancy.

Ironically, the piece of the business plan that the press found most laughable was the piece that fell into place most quickly. When Rick Tumlinson spoke at a space tourism conference about MirCorp, a former NASA scientist turned financial consultant named Dennis Tito happened to be in the audience. During the Apollo era, Tito had worked at the Jet Propulsion Laboratory plotting rocket trajectories. He parted company with NASA during its massive cutbacks in the early 1970s and went on to study finance at the University of California, Los Angeles. Forming the investment firm Wilshire Associates, Tito attacked the stock market as though it were a rocket trajectory, using computer modeling and mathematical analysis. His personal fortune went into orbit.

All the while, he maintained a strong interest in space. He attended space conferences and occasionally lectured about the importance of free enterprise in space. By the early 1990s he had accumulated a substantial fortune and the urge to fly in space. The dream didn't seem that farfetched. He had the money. Until the *Challenger* disaster, NASA had entertained plans for a robust civilians-in-space program. And now the ISS was emerging as a destination to which your average space traveler might go. He made tentative inquiries at NASA but got nowhere.

So naturally, Dennis Tito was intrigued when Rick Tumlinson spoke about the efforts of MirCorp to privatize *Mir*. Unlike the reporters at the London press conference, Tito didn't laugh when Tumlinson said that MirCorp planned to take paying customers into space. The pair met at the office of Wilshire Associates in Santa Monica. "During that meeting we looked each other in the eye and I told Mr. Tito MirCorp was real, and serious," Tumlinson recalled. "He looked at me and said 'I want to go.'" They shook hands, and the deal was done. "I called MirCorp and Anderson from the car and told them Tito was 'a go' and to contact him right away to handle deposits." Tito became the first individual in history to use his own money to pay for a space flight, MirCorp's first "citizen explorer."

Actually, MirCorp and another firm, Space Adventures, were both courting Tito. Space Adventures wanted to handle arrangements for him to fly to the ISS. Something of a tug-of-war ensued, with Spaces Adventures arranging a cosmonaut training experience for Tito and MirCorp offering him an earlier opportunity to reach space.

By June 2000 Tito was in Star City undergoing medical tests and training, while other would-be astronauts lined up behind him. Italian Carlo Vibert had signed on with MirCorp. Vibert worked for the European Space Agency and had already undergone initial *Mir* training. He planned to finance his flight with the support of a wide range of Italian industry, media, and banks. James Cameron, the Oscar-winning director of the movie *Titanic*, not only wanted to visit *Mir*, he also wanted to complete a space walk. He passed a preliminary medical screening at the Russian Institute of Medical and Biological Problems and planned to enlist various financial backers.

Unlike Tito's plan to pay for his flight out of his own pocket, the Vibert and Cameron plans relied on finding financial backers. That business model had not worked for Helen Sharman and Project Juno and would not work on this occasion either. In the end, both individuals failed to secure sufficient backers to pay the $20 million cost of the flight.

Things looked more promising, however, when Mark Burnett came along. Burnett was the producer of the highly successful reality TV show *Survivor*, and he wanted to use the same formula to build a show around a trip to *Mir*. *Destination: Mir* would open with a two-hour premiere, introducing a dozen civilians gathered at Star City, Russia, to undergo cosmonaut training. Each week one participant would be cut from the program. In the dramatic finale, the finalists would gather at the launchpad, a lucky winner would be selected, and he or she would board the Soyuz spacecraft for a trip to *Mir*. MirCorp made preliminary arrangements, and NBC was so confident that it began airing promos for the show.

The idea looked so promising, by Hollywood standards, that within a week a NASA media contractor, Dreamtime Holdings, Inc., was making the rounds at ABC, CBS, and Fox to pitch a similar show in which twenty contestants would train as astronauts at NASA's Johnson Space Center, with one finally selected to visit the ISS. Unfortunately, they neglected to clear this in advance with NASA, which quickly squelched the idea. "We're not ready for a space traveler thing," a NASA spokesman explained, "at least not over the next few years."

Given NASA's fierce resistance to MirCorp and Russia's attempts to commercialize its space program, attendees at the International Space Symposium in October 2000 might have doubted their ears when NASA administrator Dan Goldin opened the conference with his vision for what could be accomplished on the ISS. He reproached aerospace companies for not developing more visionary and entrepreneurial ways to use the space station.

"We'd like to be able to turn over the keys of the space station to a private corporation," he said, "if some of my dear NASA colleagues will have the courage to let that happen." He admonished those at NASA, "Don't be afraid of the private sector."

The private sector had long played a role in the U.S. space effort; still, this sounded like a sea change in attitude, for which MirCorp could take some credit. By pushing the envelope on commercial space, the company was changing the way that people thought about space — tourists in space,

TV shows in orbit, private investors running their own space program, a Russian rocket company leading the way in space commercialization. It had been one of the goals agreed to by Walt Anderson and Jeff Manber, and it had been accomplished. They had already changed the way that people thought about private space.

Unfortunately, the ominous clouds gathering around *Mir* in the fall of 2000 seemed to indicate that the goal of making a commercial success of the space station was proving infinitely more difficult. Conflicting news stories told a manic tale of success and failure. While MirCorp announced deals with TV producers and plans to sell stock in the corporation, Russian officials announced plans to deorbit the station.

The same month that Dan Goldin addressed the International Space Symposium, Yuri Semyonov, president of NPO Energia, and Yuri Koptev, general director of the RSA, met to consider their options for *Mir*.

"Everybody understood that there was no other choice for *Mir* but to be deorbited," Sergei Gorbunov, an RSA spokesman, reported. This understanding was, however, just a recognition of the fact that there would be no money to keep *Mir* in orbit.

"MirCorp is currently begging Energia on its knees not to deorbit the station," Gorbunov said. "They are promising to find money in three weeks, but it will be too late anyway. Spaceships cannot be manufactured in a couple of days."

To forestall the inevitable, Jeffrey Manber published an open letter to Russian president Vladimir Putin in one of Russia's major newspapers, *Kommersant*. *Mir* was the "pride and precious creation of Soviet and Russian engineers," Manber charged, "an eloquent example of Russian leadership in the field of manned spaceflight." He appealed to Putin's national pride to intervene, save *Mir*, and protect Russia's economic and scientific interests.

Commenting in response to the letter, RSA's Gorbunov played down MirCorp's role. "MirCorp's successes in *Mir*'s commercialization had been very modest, to say the least," he stated. "I just don't think Manber is competent to shape Russia's course of action in the area of its national space program."

In the first week of November, the Russian government budgeted $25 million for a January launch of a rocket to deorbit *Mir* in March 2001. Over subsequent months, Manber and the MirCorp team watched the inevitable unfold.

All of this was extremely frustrating for Manber, trying to orchestrate the MirCorp business plan. But a perfect storm of hurdles dogged the project at every turn. How do you predict in your business plan the occurrence of sun spots that expand Earth's atmosphere and hasten the decay of *Mir*'s orbit, or the U.S. government's reluctance to approve the use of the electromagnetic tether, or the infighting between Energia and the RSA, or NASA's attempts to thwart your best efforts, or brash TV producers who wanted to script out the moves of the Russian space program?

With *Mir*'s fate sealed, deals started to unravel. Mark Burnett cut MirCorp out of the loop of his plans for his *Destination: Mir* TV show and began negotiating directly with the Russians. If *Mir* came down, he was pushing to stage the production on the ISS.

Without *Mir*, work got under way to redirect Dennis Tito to the ISS. MirCorp handed over Tito's down payment to Energia. Ongoing arrangements for Tito's flight were taken over by Space Adventures. Unlike MirCorp, Space Adventures was a broker, a middleman, working with the Russians to arrange Tito's flight. MirCorp, on the other hand, had wanted to provide, not arrange for, such services. They leased the space station and underwrote the cost of service missions; in a sense, they had opened the hotel in space to which people could travel. But the hotel was closing, and bitter feelings between MirCorp and Space Adventures led to threats of lawsuits.

"Some people invited me on this cruise to watch the *Mir* come down," Jeff Manber recalled in a 2009 interview. "It's like your baby's dying, do you want to watch? I mean, I had worked on that thing commercially since '88 with the Payload Systems project. It was my whole career."

Instead, Manber opted to spend the day of *Mir*'s deorbit in his Paris apartment, consoling himself with a very expensive bottle of wine and the realization that MirCorp had very much achieved its goal of injecting free-market principles into space and laying the foundation for the entrepreneurial space companies that would follow.

5. Citizen Explorers

This isn't about Tito or safety; it's about control and
NASA's inability to grasp that space belongs to the
people, and the time has come to give it to them.

Rick Tumlinson, president, Space Frontier Foundation,
on NASA's opposition to the flight of the
first space tourist, Dennis Tito

On 27 April 2001 a chartered jet filled with Western tourists arrived from
Moscow at the Baikonur Cosmodrome. They had come to witness the
launch the next day of the world's first space tourist, Dennis Tito, aboard
the Russian spacecraft *Soyuz TM-32*. It would be one small piece of history
for the venerable launch complex but a groundbreaking event in the fledgling business of sending private citizens into space.

Baikonur is the largest and oldest space launch facility in the world. *Sputnik*, the first Earth-orbiting satellite, had begun its journey here in 1957, as
had Yuri Gagarin, the first man in space, in 1961. All subsequent Russian
manned missions had launched from here, as well. Baikonur was a symbol
of the storied history of the Russian space program.

However, in the post-Soviet era, Russia no longer owned its premier rocket
launch facility but instead leased it for $115 million a year from the now in-
dependent nation of Kazakhstan. That change of ownership was but one
small measure of how much the space landscape had shifted in Russia since
the fall of the Soviet Union.

The visitors who emerged from the chartered jet were symbolic of another shift. They were "space tourists," a new demographic that had begun to infuse much-needed revenue into the cash-strapped Russian space
program. Working with Western partners, Russia had started to cater to

those who wanted a space experience—a tour, a cosmonaut training camp, a zero-g airplane flight. Dennis Tito took the formula to a whole new dimension. For a reported $20 million, he bought a ticket to the International Space Station (ISS).

Twenty-five of Tito's family members and friends had traveled to Russia for the launch. His ex-wife and business partner, Suzanne, was among them, as were his sons, Michael and Brad; his girlfriend, Dawn Abraham; and several wealthy friends and colleagues from his investment company, Wilshire Associates.

Accompanying Tito's group was the staff of Space Adventures (SA), a pioneering space tourism company headquartered in Arlington, Virginia. SA CEO and president Eric Anderson stepped off the plane, along with a handful of SA staff and investors. Though his prematurely receding hairline suggested otherwise, Anderson was a mere twenty-six years old, but he was about to establish himself as the primary player in space tourism.

Anderson had brokered the deal to fly Tito to the ISS and had survived through prolonged bargaining with the Russians and strenuous NASA opposition to the flight. Over the past few years, by patiently nurturing complex agreements and adding Russians to his staff, he had come to understand the Russian way of doing business. He had learned how to package the space experience. But Dennis Tito was Anderson's first orbital client, and his flight would have a huge impact. Media attention was intense. The flight would redefine not only Anderson's business but the very concept of access to space.

Having disembarked from the aircraft, the group was herded aboard two minibuses for the hour-long ride over bumpy roads to lodgings at the Sputnik Hotel. The bus rolled past clusters of gray buildings, endless stretches of desert, and the occasional camel wandering in the distance. Baikonur is vast, about the size of the state of Delaware. Built during the cold war for launching intercontinental ballistic missiles (ICBMs), the whole complex looked as though it had been plunked down in the middle of nowhere and all but forgotten—which was not far from the truth.

Was it permitted to take photographs? someone on the bus asked the Kazakh tour guide. "Sure," she replied. "But there is nothing interesting to take a photo of." Moments later the bus drove by discarded N-1 rocket tanks, the legendary boosters that would have carried the Russians to the moon and that were now being used by locals as garden sheds. It was but one more reminder that the group had ventured into a time warp between the glory days of the Soviet space program and the brave new world of space tourism they were helping to create.

The group settled into the Sputnik Hotel. With its landscaped lawn and freshly painted yellow façade, the hotel looked as though it had been transported from a European resort. Owned and operated by the French company Starsem and dubbed "an oasis in the desert," it was the only Western establishment in the city. It catered to the occasional foreigners who visited Baikonur for commercial unmanned launches. The Cosmonaut Hotel, Tito's quarantine quarters, stood next door to the Sputnik. His family would have a chance to visit him there later that day.

In the middle of the night before launch, the phone rang in the Sputnik Hotel room of Tito's sons, Brad and Michael. The *New York Times* was calling. They were writing an article about what type of man Dennis Tito was. Media interest had grown around Tito for nearly a year and had reached full boil over the past month as NASA tried to derail his visit to the space station. The celebrity of this multimillionaire, would-be space traveler now rivaled that of any of NASA's astronauts. Interest in the ISS had never been so high.

By talking to Tito's friends, the *Times* reporter had already pieced together a snapshot of the man. He loved sailing, opera, and fast cars, although friends claimed he drove the cars slowly. Los Angeles mayor Richard Riordan had added, "I've known Dennis for thirty years, and he's never grown up. He is just always looking for exciting things to do."

But now the reporter had Tito's own sons on the line, and he wanted to know what really drove Tito and what kind of father he was.

"The way I see it, the guy grew up in Queens and he wasn't well off. He went to school and worked for NASA, did some pioneering work there, and went off to make a lot of money," was how Brad Tito tried to sum up his father. "Basically, he's at the point in his life where he's thinking about what's really important, and in the end, when we all ask ourselves that, it ends up being life experience."

Michael Tito, who had joined his father in Russia the previous year for a zero-g flight and centrifuge training, said his father had always taught them to "think outside the box."

While these appraisals of Tito and his motivations may have been accurate in the past, they missed the change that had overtaken him during preparations for his flight. In interviews leading up to his launch, he tried to explain what he now thought about his role. "Originally, it was just a desire to go into space," Tito told *CBS News*. "I really think I didn't understand the broader implications. And over the last nine months, I realized that it is something unique and pioneering. I'm actually getting more and more excited about the future as far as opening up space to private citizens, developing the enthusiasm for other people to follow in my footsteps, and open up space to the general public."

Tito's preparation had begun the previous June, when he took up residence at the cosmonaut training center at Star City, near Moscow. The actual Gagarin Training Center, the core area for his training, lay within a gated compound. Soviet-era buildings dotted the grounds, each guarded at the entrance by a lone sentry. Tito resided at the foreign cosmonaut quarters, within a three-story building that looked like a 1960s apartment complex that time forgot.

His spare one-bedroom apartment was in stark contrast to his thirty-thousand-square-foot Pacific Palisades mansion, rumored to be the biggest private residence in the LA area. (It was featured in the movie *Wag the Dog*.) Aside from his laptop computer and printer, his simple quarters included no hint of luxuries. It was Tito's home for most of his eight months of training.

Being a millionaire gave Tito no slack during training. "The Russians didn't cut any corners," he boasted about his training. He had to pass the same medical, physical, and technical milestones as any other cosmonaut in training, along with the additional burden of learning the Russian language.

From June to August 2000, Tito stayed in Star City for periods of several weeks at a time and trained individually with an interpreter on the theory of cosmonautics, navigation, and aviation — material familiar to him from his days at NASA. He learned about the intricacies of the Soyuz TM spacecraft, his home for two days between launch and docking, as well as the Soyuz launch vehicle that would transport him off the planet.

At this point, he was still scheduled to visit the *Mir* space station.

Although he would have no specific responsibilities during the mission, he had to know how to perform some critical and emergency procedures. This was achieved using full-scale Soyuz simulators, able to replicate any conceivable spaceflight scenario. On parabolic flights on the Ilyushin-76 aircraft, he learned how to move in weightlessness, pass equipment to crewmates, and get in and out of his Sokol suit. Later, when his destination switched to the ISS, he needed hundreds of additional hours of training to become familiar with the Russian segment of that station in a full-scale ISS mockup.

Although Tito was physically fit for a fifty-nine-year-old, running and daily weight training and cardio sessions readied his body for the rigors of spaceflight. In preparation for the maximum 8-g forces he would experience on the descent phase of the mission, he trained in the TsF-18 Centrifuge, the biggest in the world.

At certain milestones of his training, medical and health examinations gauged his fitness to continue. Medical certifications were critical to the success of the mission and were not taken lightly by the now renamed Russian Aviation and Space Agency (RASA). If he failed any of them, he would be out of the program.

Tito claimed his training in Moscow was "a lot like going to college during one of those semesters when you have to buckle down and get good grades." Despite the challenge, he bore his circumstances well. "I'm just a recruit," he said. He would later reveal that the hardest part of training was not knowing whether he was actually going to fly. *Mir*'s fate was uncertain, and the feasibility of a flight to the ISS seemed far in the distance.

In the fall of 2000, when MirCorp failed to raise sufficient funds to pay for a service mission to the station, *Mir*'s fate was sealed, and Tito shifted his attention to arranging a trip to the ISS. "I don't believe the Russians are eager to default on my contract. There is a considerable amount of money involved, and the Russians are much more savvy these days about world opinion," he said in a Space.com interview. Still, Tito would not admit to being certain about his flight until he was "bolted inside that capsule at Baikonur." A year of dealing with the Russians and with the fledgling space tourism industry had made Tito wary of promises.

The Russians may have been savvy about world opinion, but not about the storm that would erupt following their decision to allow Tito to change his flight plans from *Mir* to the ISS. For the Russians, it was a simple calculation. As a major ISS partner, Russia provided unmanned Progress vehicle resupply missions and Soyuz crew transports to the station. Although the Soyuz had three seats, it only required a crew of two, a commander and a flight engineer. In Russia's view, since it was required to provide these missions at its own expense, it did not need the permission of ISS partners to select a crew, including the choice of Dennis Tito. The other ISS partners saw it differently.

Although the memorandum of understanding between NASA and its ISS partners did not address civilian or commercial visitors to the station, it did stipulate that consensus must be obtained between partners with regard to safety and crew-flight opportunities. The European Space Agency (ESA) station chief, Jorg E. Feustel-Buechl, condemned Russia's plans as "irresponsible."

After a hurried series of meetings in Houston, Cape Canaveral, and Moscow showed unanimous objections from all ISS partners, NASA administrator Daniel Goldin explained the space agency's stand on Tito's flight. NASA supported space tourism to the ISS, he stressed, but the timing of this flight was wrong. Too much work remained to be done on the station, and the presence of a nonprofessional crew member could potentially put the mission at risk. Golden further stated that a consensus had to be reached by all ISS partners concerning any use of the station for commercial purposes.

Despite this opposition, ISS's Multilateral Coordination Board continued to work on a document that would officially be titled "Principles Regarding Processes and Criteria for Selection, Assignment, Training and Certification of ISS Expedition and Visiting Crewmembers." Although no established process existed for regulating Tito's flight, such a document had been under development for over two years, overseen by astronaut Charles Precourt at NASA's Johnson Space Center (JSC). However, since Tito's flight came before these space traveler guidelines were finalized, the ISS partners voted against letting tourists onboard the space station.

Unfortunately for the partners, they had no way of imposing their decision on the Russians. At that time, the station was composed mostly of Russian modules, including the crew quarters. As NASA senior station manager Mike Hawes confessed to the media in early February 2000, "The Russians have the key to their spaceship, so from a physical point of view,

I don't have any way of keeping the Russians from launching a rocket with somebody else in it."

Even as the dispute simmered between NASA and RASA, Tito prepared for his mission. He announced to the press the signing of his ISS contract with RASA, at a reported cost of $20 million for a ten-day flight. "I'm really excited, I mean, I just can't wait to get in my seat and go," he said in a Space. com interview. "I'm going to be doing nothing else for the next three months. It's just going to be a full eight hours a day of training." Tito was already back training in Star City and worrying that health problems might sabotage his dream. In January he had been hospitalized with a bout of pneumonia. He had to be extra vigilant in Moscow's harsh winter weather not to allow a cold to cause a medical disqualification.

The strained relationship between RASA and NASA finally hit bottom on 19 March, when Tito and his Russian crewmates, together with their backups, arrived at JSC to fulfill the forty-hour cross-training requirement. In accordance with Russia's partner agreement, any crew member headed to the ISS needed to cross-train for a week at NASA JSC to become familiar with the American segments of the ISS. Tito was not expected to enter the U.S. segments of the ISS, but the Russians brought him anyway to fulfill the agreement. NASA had already made clear its opposition to Tito's visit. With a clear sense that a confrontation was brewing, *Newsweek* magazine sent a reporter to chronicle the entire event.

When the Russian group arrived at JSC, they first huddled in the parking lot. Commander Talgat Musabayev conferred in Russian with his fellow cosmonauts, then addressed Tito in English, using his nickname, "Denny, we are together. We have to be staying together. Understand?" Tito repeated the instructions, and they all slapped each other on the back.

Inside, the impasse came quickly. Bob Cabana, a former astronaut and director of NASA's International Space office, led the group into a conference room. Cabana told Tito, "The NASA position is still we do not think there is adequate time to train you for an April flight. But regardless of when you fly there are technical issues to settle regarding reimbursements for training, insurance issues, medical issues. So we'd like to sit and talk with you about that this morning."

Musabayev explained that Tito had already trained for seven hundred hours in Russia and that the group must do training together.

"In that case, we will not be able to begin training," Cabana responded, "because we are not willing to train with Dennis Tito."

Musabayev insisted they all had "the same knowledge of the American segment. We start from zero. It is the same for Mr. Tito." But Cabana remained firm. "There is nothing to discuss; we have given you our position." The meeting ended and the entire crew walked out of JSC together.

The following day, NASA released a statement and held a press conference regarding the incident. NASA, it said, "supports the commercialization of the International Space Station . . . however, based on incomplete crew criteria and unresolved operational and legal considerations, there is not enough time to prepare Tito for a safe Soyuz flight to the station in April." Station manager Mike Hawes claimed that Tito would need to train some eight weeks at JSC, even though this training typically took only one week for any Soyuz professional crew visiting the station.

Despite repeated assurances from Hawes during the press conference that the whole situation was a "partnership issue" and that it would be resolved between all the ISS partner states, the Russians showed no sign of backing down on their promise to fly Tito in April. Yuri Koptev, head of RASA, told the Interfax news agency that Tito would fly to the station "regardless of the position of foreign partners."

The next day, Tito's crewmates returned to JSC without him, following directives from Moscow. However, they assured Tito that all would be well. "You will fly with us," said mission engineer Yuri Baturin. "You are member of crew, I am sure of this."

Waiting back at the hotel, Tito received encouraging phone calls from his aerospace lawyer and from his friend, Apollo astronaut Buzz Aldrin. His spirits were further buoyed when he treated the cosmonauts to a lunch at a restaurant on the Texas gulf coast. Hoisting a pricy glass of Sonoma Valley chardonnay, Musabayev gave the toast, "To our mission." Tito reciprocated with his own toast: "We're going to make history." Musabayev beamed. "History, I don't know, but we need the money."

Despite continued opposition to Tito's flight by NASA and its ISS partners, the would-be space tourist returned to Moscow for two more weeks of formal training, where he passed all his medical requirements and headed off to Baikonur on 16 April.

Even at this point, Tito still did not know whether he would be allowed to fly to the ISS. However, Russian determination left NASA with no room to maneuver. Five days later, the space agency finally dropped its opposition after obtaining a signed agreement from Tito releasing NASA of any liability in case something went wrong on his flight. Tito also promised to pay for any damages he might cause to the multibillion-dollar station. He would also have only limited access to non-Russian segments of the station, and station activity would be kept at minimal levels because of his presence.

Four days before launch, the Multilateral Coordination Board, comprising all ISS partners, officially granted Tito a one-time-only exemption to fly to the ISS. However, that one action did not clear the air of NASA's resentment. Only the previous month, the agency had felt compelled to force Russia to deorbit its aging *Mir* space station so that Russia could focus its resources on the construction of the ISS. Now it had knocked heads again with its space partner because some rich guy had bought his way onto the space station. It was a troubling pattern.

Questioned by Space.com senior space reporter Leonard David on the day of Tito's launch, NASA administrator Daniel Goldin gave a blunt response: "Space is dangerous. It's not about a joyride. Space is not about egos."

Tito was already suited up in his white Sokol space suit behind the glass wall, answering questions from a room full of journalists and space engineers, when his entourage arrived for the preflight press conference. His girlfriend, Dawn Abraham, blew him a kiss through the glass partition. She would later confess to MSNBC that Tito had wooed her with talk about space. "That's why I started to like him." Also in attendance to watch their former MirCorp client were Jeff Manber and Walt Anderson. After the press conference and suit check, Tito left the building with his two cosmonaut crewmates, Musabayev and Baturin.

The men walked onto the tarmac and took up positions on painted squares that represented the seat positions they would occupy in the Soyuz. Then Commander Musabayev reported to Pyotr Klimuk, chief of the Gagarin Cosmonaut Training Center, and to Yuri Semyonov, head of Energia Rocket and Space Corporation, the giant Russian aerospace company. The crew was ready for their mission, he told them. Looking on were such luminaries as cosmonauts Vladimir Titov and Valentina Tereshkova, the first woman in space. As the crew boarded the customized RV transport, the crowd scrambled into vehicles to follow them to the launchpad.

When the RV stopped on the way to the launchpad, the following vehicles waited at a discreet distance. Here the crew made the traditional "pit stop"—urinating on the wheel of the cosmonaut van—said to have been pioneered by Gagarin on the way to the launchpad in 1961. It had become a requirement for every manned launch at Baikonur since then.

Ironically, the quarantine for the crew, which was strictly followed during the press conference and suit-up, was forgotten as the space-suited cosmonauts, at this stage minus their helmets, wove their way through the crowd to get to the foot of the launchpad. For those in the group who had experienced a NASA launch, it was amazing to see so many people milling around the launch vehicle only three hours prior to liftoff. The rocket felt alive, with the atmosphere freezing on the side of its cryogenic oxygen tanks and snowing down on the crowd.

The new age of space tourism seemed on full display as everyone busily snapped souvenir photos in the shadow of the Soyuz rocket. An official translator reminded the group not to smoke, waving his hand in the direction of the venting rocket fifty feet away, filled with ninety tons of rocket fuel. A large sign for the Russian Space Agency (Rosviacosmos) division for general construction in Baikonur—called KBOM—was read by the Americans as "kaboom."

When the crew ascended the elevator to the top of the rocket, guests adjourned to the launch viewing site, a few covered structures less than a kilometer away. These were surprisingly close to the action, compared to the Kennedy Space Center's seven-mile restricted zone for visitors viewing shuttle launches. There was no visible security apart from the chest-high metal fences that reminded people not to cross the no-man's-land between them and the rocket.

Unlike shuttle launches, which have specific countdowns broadcast for the visitors, Soyuz launches have none. A short announcement in Russian from the overhead speaker and the swing of the gantry arms away from the launchpad were the only indications before the Soyuz rocket lifted off.

A spontaneous cheer erupted in the launch stands from Tito's friends and family and the staff of SA. A cockpit camera onboard the Soyuz spacecraft recorded Tito waving good-bye through the right seat portal, through which he would finally realize his dream of seeing Earth from space.

At mission control, the Russians were also applauding, including eighty-nine-year-old Boris Chertok. Chertok had been involved with the Russian space program as long as there had been a Russian space program. He had seen it all. He had worked as a designer for the legendary Sergei Korolev. During the cold war, his guidance and control system went into the first ICBM, the R-7. During the U.S.-USSR space race, he worked on the Vostok, Voskhod, and Soyuz spacecrafts.

Now Chertok had just witnessed a Russian rocket launch a paying American customer into space. He labeled the occasion "a very important step in showing that regular people can fly to space. . . . I am a little bit under ninety, and if someone presented me with $20 million, I would gladly spend it on a ticket to space."

Yuri Grigoryev, deputy chief designer for the Russian rocket-builder Energia, put it more succinctly: "The station is open for commercialization."

Back at the viewing area, the crowd followed the rocket upward until it was a barely visible speck. A few people cheered and others fought back tears when the rocket finally disappeared from sight. Nine minutes after liftoff, Tito would be in zero g and traveling at over nineteen thousand feet per second. In another ninety minutes he would have circumnavigated the Earth. He would, on this date, 28 April 2001, become the 404th person to fly in space and would make history as the world's first space tourist. Hundreds of cosmonauts had lifted into space from this facility, men and women, foreign cosmonauts, engineers, doctors, and a Japanese journalist, but before Dennis Tito, not one of them had ever purchased their own ticket to space,

No one cheered louder for the successful launch than the unassuming, blue-jean-clad figure of Eric Anderson, who deserved much of the credit for having just launched the new industry of space tourism.

Eric Anderson began thinking seriously about space tourism in the mid-1990s while still attending the University of Virginia. He was studying aerospace engineering and nurturing a desire to travel to space himself. However, during a much-prized summer internship at NASA, he came to two important conclusions: his 20/400 eyesight would prevent him from becoming a NASA astronaut, and NASA bureaucrats would not be bringing down the cost of space travel in the foreseeable future.

"Things there are massively overinflated," Anderson concluded. "I'd see a million-dollar study produce a 100-page report that I could have written in college. The government has no incentive to make things cheaper. The bigger the budget, the more power they wield."

Fresh out of college, at the tender age of twenty-three, Anderson cofounded Space Adventures with a group of space and travel pioneers, including Peter Diamandis from the X PRIZE; Mike McDowell, whose companies Quark Expeditions and Deep Ocean Expeditions had pioneered high-end adventure travel; and Gloria Bohan, president and CEO of Omega World Travel, the largest privately owned tour agency in the United States.

In the early years, SA ran like many of the quintessential garage dot-com companies of the 1980s and '90s. Anderson filled the company's headquarters—in the basement of his Arlington, Virginia, townhouse—with equally star-struck employees enthusiastic about the company's goals. The staff, many in their twenties, worked long hours to cover customer service, administration work, marketing, program development, and operations. The company's energizing vision was that people would pay money to fly in space.

"Space Adventures was like a college project," according to Robert Pearlman, SA's first employee and director of public relations and marketing. "It felt like we were tackling a much bigger project than either of us could do." According to Pearlman, everyone experienced on-the-job training. People were given tasks and expected to run with them with little guidance or supervision, from making high-level marketing deals to creating new programs in the shortest possible timeframe. In retrospect, the employees' gung-ho attitude and lack of experience worked to their advantage, because they didn't know any better and thought everything was possible.

Anderson himself had a type A personality—energized, highly impatient, and quick to recognize opportunities. His cell phone would permanently attach to his ear while he multitasked on his computer or while driving through traffic. He was a workaholic and expected the same dedication from his subordinates. He would jokingly say, "Are you taking a half day off?" if an employee left by 6:30 p.m., yet would eagerly encourage them to go watch the premiere of *Star Wars* in the middle of the workday.

Although SA could not yet deliver space to its customers, it could most definitely deliver thrilling space-themed adventures.

SA developed a repertory of programs trademarked as "Steps to Space." These ranged from flying in supersonic MiG jets in Russia, watching a space shuttle launch at Kennedy Space Center, to hunting for meteorites in Antarctica. The general concept was to offer a range of space-related experiences to space enthusiasts until the day they could finally fly in either a suborbital or orbital vehicle.

With the support of a few angel investors and the success of the company's tour operations, SA's client list swelled with well-heeled adrenalin junkies eager for the next adventure to tick off their list. Want to hop over to Russia's Star City cosmonaut training facility and fly zero-g flights with genuine cosmonauts? SA could arrange it for $5,000. Which is what they did for the likes of film director James Cameron and fashion designer Isabella Kristensen. Anderson would often accompany his notable clients to Moscow and Star City to ensure VIP treatment.

One of SA's early investors and frequent clients was Richard Garriott, a multimillionaire computer game developer, entrepreneur, creator of the Ultima series of computer games, and the son of retired NASA astronaut Owen Garriott. He would later fly as Anderson's sixth orbital client.

Anderson met Garriott while manning the SA booth at a New York Explorers Club gala. Garriott stopped by the booth and asked Anderson, "Are you trying to get people to space? I want to go." That opening led to Garriott investing much-needed seed money that sustained SA in the early years.

In many respects, the involvement of Richard Garriott in the space tourism industry was the most natural of connections. His willingness to invest in SA fit with his desire to fly into space himself. "For me, space exploration was tangible; it was part of my family's DNA," he explained on his space mission blog. He had always wanted to follow in his father's footsteps, but he knew his chance of getting to space through NASA was pretty slim. "There was no imaginable way I was going to get a front-row seat like my father. Even he, who had the perfect background for NASA, was still incredibly lucky to be selected over the thousands of other worthy applicants. I knew that if I wanted to go, it would have to be on some non-existent private manned space program."

Garriott nurtured his space interest through his investments. "I backed my father's early attempts to privatize aspects of the space program through a company we called EFFORT, which tried to extend shuttle missions with extra fuel carried in the cargo bay," Garriott explained. "I was also one of the earliest investors in SpaceHab, which now flies a variety of space hardware in shuttles and to the International Space Station." Later he would back such space projects as the X PRIZE Foundation and Zero-G Corporation, founded by one of his father's crewmates on STS-9, Byron Lichtenberg.

Garriott's passion for space extended to other adventure pursuits, including tracking gorillas in Rwanda, searching for meteorites in Antarctica, canoeing down the Amazon, diving to the *Titanic*, and going on hydrothermal vent expeditions. So it was really no coincidence that Anderson met Garriott at the New York Explorers Club or that Garriott agreed to underwrite SA's business plan that called for people to buy trips to space.

In the early years, as SA built its business, it struggled to penetrate the Russian bureaucracy. Anderson worked hard to find the right officials in Moscow and Star City to allow access to facilities for paying customers. Once he managed to make arrangements to offer space-related experiences, he had to determine how best to package the experience for his affluent clients. Although luxury hotels and services were beginning to appear in Moscow, at nearby Star City, the center of Russian space activities, the time-worn facilities had the feel of a vintage military base.

And yet surprisingly, SA clients, who paid premium prices for these adventures, quickly adjusted to the austere amenities and environment. That spartan atmosphere became part of the adventure package, the sacrifice that clients were prepared to make to venture into this exotic post–cold war culture and be a cosmonaut for a day. So they endured with patience long lines at airport immigration, no bathroom stops on long transfers to the airbases, and no snack shops or soda machines at any of the training facilities. For them, the drab, big-brother-is-watching atmosphere at Star City and Zhukovsky Air Base took on a James Bond flavor. The exotic language and the super-secret Russian destinations appealed to clients seeking adventure.

SA built its Russian business on MiG flights and cosmonaut training, but from the outset, Anderson had been convinced that a market existed for sending tourists into space. He once toyed with the idea of using the space shuttle for this purpose, but abandoned the idea after initial talks with NASA made it clear that paying passengers would not be allowed to fly on the government-funded shuttle.

By 1998, on the strength of the business he had developed at Star City, Anderson began talking with NPO Energia about the possibility of putting passengers on orbital flights. With Helen Sharman and Toyohiro Akiyama setting the precedent by flying to *Mir*, Anderson thought it would be possible to find a way for private citizens to visit the ISS.

Anderson told the Russians what he thought would be practical for his clients in terms of cost and time commitment, and commissioned them to conduct a feasibility study on whether orbital flights could be delivered. The study concluded that the best approach would be to put a paying passenger in the empty third seat on the ISS Soyuz taxi missions. Taxi missions were normally scheduled twice a year to replace the required Soyuz escape vehicle attached to the ISS.

That was the package Anderson had been looking for. Working within the regular schedule of Soyuz flights, it would be possible to fully train private citizens to take on the minimal responsibilities of the third flyer and visit the station for the few days it would take the Soyuz crew to do their housekeeping and switch spacecraft before returning to Earth.

However, the dollar figure the Russians came up with for each seat opportunity appeared to be a considerable hurdle—$20 million. No one at that time could imagine that anybody would pay that amount for six months of preflight training and a wild space ride.

SA investor Richard Garriott was one person willing to do just that. At the time, Garriott wanted very much to be the first person to take advantage of this opportunity and to follow his father's career as a space flyer. Unfortunately, his personal finances suffered during the dot-com crash, forcing him to postpone that dream. He would, however, eventually fly to the ISS in October 2008. But for now, Anderson had to find another client dedicated enough to want to be a space pioneer who also had the funds to make it happen.

In early 2000 a newspaper article appeared on Eric Anderson's desk. An investment executive named Dennis Tito had mentioned in an interview that he wanted to fly into space. Anderson set up an appointment to meet with Tito at his LA office. He went to the meeting with the Russian feasibility study in hand, hoping to get the fifty-nine-year-old multimillionaire to invest in SA. What he got instead was a client eager to occupy that empty Soyuz seat to the ISS.

Within a few weeks, Tito and Anderson had negotiated and signed a contract wherein SA would work to provide Tito with an orbital space flight opportunity to the ISS and Tito would agree to fly if such an opportunity was provided. Anderson convinced Tito to travel to Moscow to meet with the Russians and to proceed with initial medical exams and pretraining activities. This would give him a taste of the six months of training required for an orbital flight.

Tito's SA package included a centrifuge test, Soyuz simulator training, and a zero-g flight on the Ilyushin-76, the Russian equivalent of NASA's "vomit comet." Topping off the whole experience would be a supersonic flight in a MiG-25 fighter jet.

The MiG-25 Foxbat could fly straight up at speeds more than two and a half times the speed of sound to over eighty thousand feet. It was about as close to a rocket flight as you could get without actually being in a rocket. At the conclusion of his training experience, Tito strapped into the MiG-25 and got blasted to a height more than three times the cruising altitude of a commercial airliner and above 99 percent of the planet's atmosphere. There he saw the curvature of the Earth and the blackness of space above him. Below him extended a horizon more than seven hundred miles wide. No sooner had the MiG reached its highest point than it began its maneuver to return to Earth. Tito had almost reached space. How could he now pass up this opportunity to experience the real thing?

What only became apparent at that point was that Tito had not one but two options for achieving spaceflight. Two Russian astronauts had just returned from a seventy-three-day mission to *Mir* during which they had repaired air leaks, replaced deteriorating batteries, and revitalized the aging station's life support systems. The renovated *Mir* had been declared ready for commercial flights.

The Russians informed Tito that he might be able to get to space sooner if he was willing to attempt a flight to the aging *Mir* space station through the commercial company MirCorp, as opposed to a flight to the ISS. Intent on being the first tourist in space, Tito agreed to the earlier opportunity to go to *Mir*.

It proved to be only a temporary detour from his ISS plans. *Mir*'s troubles quickly mounted, and conflicts over Tito's payments to MirCorp finally led Tito to withdraw from the agreement. Eric Anderson and SA reentered the picture to handle arrangements for the shift to an April 2001 flight to the ISS. It had been a long hard slog, but a short four years after establishing SA, Anderson had managed to book his first client on an orbital flight.

After Dennis Tito's launch, Eric Anderson sat in the lobby of the Sputnik Hotel looking exhausted and subdued. Everyone had been through so much to get to that point, to launch an industry. The moment seemed to beg ceremony. While one SA staffer rolled a video camera, Anderson recorded his thoughts. "It was a huge day. April 28, 2001. First space tourist flies to space. Long time in the making. It's overdue. We'll look back on it with a lot of pride."

The path forward from that moment seemed clear. Anderson would find more orbital clients and buy more orbital flight opportunities from the Russians. And some day he would be flying clients on privately owned and manufactured orbital vehicles, flying to privately owned space stations in Earth orbit and beyond. "I'm very hopeful that it will happen but I'm not sure when it will be. It could be ten years still. Actually, if it was in ten years, that wouldn't be bad."

Even while Tito was still enjoying his time on the ISS, his mission was already having an impact back on Earth. Other would-be space travelers were lining up to follow his lead. "The eyes of the world are on Tito now," Eric Anderson told Space.com. "It's a gateway. We've had a lot of interest from potential people who want to do the same sort of thing. There is opportunity with Soyuz taxi missions, a couple of opportunities per year for Dennis Tito–type flights."

Tito returned from his eight-day adventure to a hero's welcome. The moment he emerged from the *Soyuz TM-31* capsule on the steppes of Kazakhstan, he gushed about his experience, "It was paradise. I just came back from paradise." A helicopter whisked him to the Kazakh capital for a warm welcome and a hug from Kazakhstan's president, Nursultan Nazarbayev. "Until recently, you would only read in science fiction that an ordinary man could go to space," Nazarbayev enthused. "You have paved the way for space tourism."

Ceremonies continued back at the Gagarin Cosmonaut Training Center as space officials, politicians, and the Star City orchestra greeted the returning crew. Deputy director of Rosviacosmos Valery Alaverdov called Tito a persistent, courageous man who pursued his dream, and he praised the reliability of Russian space hardware and medicine that could support the flight of amateur cosmonauts.

With this successful mission, the age of space tourism had clearly dawned, and Dennis Tito quickly became one of its most ardent promoters. Within a week of his return, he appeared on the Larry King and David Letterman TV programs. Then it was off to Washington DC for a Space Policy Roundtable on Human Space Flight, where he claimed that the flight had been so valuable that "if they told me ahead of time that it would take all the money I have, I'd pay it and live the rest of my life on welfare."

When asked if he had a postflight message for NASA, he merely conceded that if NASA hadn't put the ISS up there in the first place, he would not have had this wonderful experience. Other speakers were not as gracious toward NASA. The man who oversaw NASA's Teacher in Space program, former NASA associate administrator Alan Ladwig, had sharp words for the agency for not using Tito in its activities. "NASA did not use this as an opportunity. Instead of bitching about things, they should have enlisted Tito to tell people about what the ISS was going to do. At the time, the only news that was being heard about the ISS was a $4 billion cost overrun."

The chorus of criticism over NASA's handling of the flight continued to rumble for several months, but the agency had already begun to adjust its attitude to the new reality of space tourism.

Subsequent clients who stepped forward for an orbital opportunity proved to be a unique breed. Each of these space tourists, or "space flight participants" and "private space explorers," as they preferred to be called, was a pioneer in his or her own right. They had already shown a certain daringness and bravado in their private lives, and they brought that same mix to pioneering space tourism. "We should be thankful," said Eric Anderson. "If there weren't people who are willing to step up to the plate and be willing to take the risk of spaceflight, and invest their own money, time, energy and capital to show that this is a market place, then the rest of us don't have a job. No investment would be flowing into the industry."

One year after Tito's flight, when South African Mark Shuttleworth visited the ISS, the criteria for civilian space flight participants were in place, and NASA welcomed with open arms his participation on the *Soyuz TM-34* mission.

The role of the participants had also begun to change. Unlike Tito, who spent most of his time in orbit viewing the Earth from his ISS window and listening to opera, subsequent flyers would front-load their flights with scientific experiments and activities to further distance themselves from the perceived image of the "joyride" tourist.

SA collaborated with both RASA and the ESA to match orbital clients with experiments that had been waiting for a flight opportunity. On other occasions, experimental flight opportunities were created for schools and educational institutions. During his flight, Shuttleworth worked on experiments dealing with the human genome and HIV virus crystallization. The HIV study addressed an issue particularly important to Africa, where the disease has been devastating.

Shuttleworth was the first person from Africa to fly in space, and he saw himself as an unofficial representative of his country and the continent. His flight aroused a national pride, and he would conduct in-flight phone calls with South African president Thabo Mbeki and former president Nelson Mandela, but he was also aware of how his experience could inspire all of Africa.

When *BBC News Online* interviewed him from space, Shuttleworth stressed how important it was for Africa to embrace its future and create a sense of excitement for the people. "One of the things I hoped to do by fulfilling my own dream was to do it in a way that might reach out to particularly children and learners in Africa and show them that dreams can come true, and that's a very powerful thought. The world changes far faster than we imagine it—the mere fact that I am sitting in the Russian segment of the International Space Station and if I move thirty meters that way I'll be in

1. Gerard O'Neill with a photograph of his Island One space colony concept, the Bernal sphere. Courtesy of Tasha O'Neill.

2. Cutaway view of a Stanford Torus. This colony design had the shape of a 1.8-kilometer-diameter donut. The rotation of the torus would provide Earth-normal gravity on the inside. Painting by Rick Guidice, courtesy of NASA Ames Research Center.

3. Would-be astronaut Fell Peters testing the cramped nose cone of the Volksrocket. Robert Truax stands behind. Courtesy of Fell Peters.

4. The power end of the Volksrocket consisted of these Atlas rocket vernier engines. Regeneratively cooled, with thrust chamber gimbaling, they were originally manufactured by Rocketdyne at a cost of about $70,000 each. Truax picked them up in a Southern California junkyard for $25 apiece. Courtesy of Fell Peters.

5. A proud Robert Truax standing before his x-3 rocket just prior to its static test firing on Media Day, 24 June 1980. Photo by Mark Crosse.

6. The Voskhod space crew, arguably the first noncosmonaut, private citizens to travel into space. Left to right, Konstantin Feoktistov (engineer), Boris Yegorov (physician), and Vladimir Komarov. RIA Novosti.

7. Teacher in Space Christa McAuliffe on the KC-135 for zero-g training.
Courtesy of NASA Johnson Space Center.

8. STS 51-D payload specialists train in JSC's shuttle mockup and integration laboratory. Senator E. J. (Jake) Garn, right, holds a color chart while Charles Walker, representing McDonnell Douglas, takes pictures. Dr. Thomas Moore, a NASA physician researcher, looks on. Courtesy of NASA Johnson Space Center.

9. U.S. representative Bill Nelson, STS 61-C payload specialist, preparing to eat a freshly peeled grapefruit on the middeck of the space shuttle *Columbia*. Courtesy of NASA Johnson Space Center.

10. The *Soyuz TM-10* space crew. Left to right, Captain Colonel Viktor Afanasyev, Japanese journalist Toyohira Akiyama, and onboard engineer Musa Manarov. RIA Novosti.

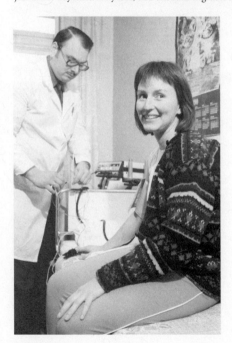

11. British astronaut Helen Sharman during a medical examination at the Institute for Medical and Biological Problems. RIA Novosti.

12. Cosmonaut Alexei Leonov, right, seeing off the *Soyuz TM-12* crew to the Baikonur space center. Left to right, captain Anatoly Artsibarsky, British researcher Helen Sharman, and engineer Sergei Krikalev. RIA Novosti.

13. Walt Anderson, founder of MirCorp, standing in front of a Soyuz capsule in the museum of NPO Energia outside Moscow. Jacobson/Gardellini.

14. Negotiating the MirCorp lease. On the Russian side (left) can be seen Alexander Derechin, Victor Legostaev, Yuri Semyonov, a translator, and Valery Ryumin. On the American side are Jeffrey Manber, Gus Gardellini, and Walt Anderson. Not seen are John Jacobson and Rick Tumlinson. Jacobson/Gardellini.

15. Dennis Tito (left) with crewmates Talgat Musabayev (center) and Yuri Baturin (right) prior to launch at the Baikonur Cosmodrome, 28 April 2001. Courtesy of Space Adventures, Ltd.

16. Every private-paying space tourist to date from Dennis Tito to Guy La Liberte has launched on a Soyuz rocket. The same pad used for Yuri Gagarin in 1961 is still in use today to launch crews to the International Space Station. Courtesy of Space Adventures, Ltd.

17. Anousheh Ansari, the first female and fifth space tourist, posing for the camera in a full Sokol spacesuit. Courtesy of Space Adventures, Ltd.

18. Eric Anderson (center) joins client Richard Garriott (in space suit) on a parabolic flight run by Zero-G Corporation. Courtesy of Space Adventures, Ltd.

19. Gary Hudson standing next to Rotary Rocket's Roton ATV. Courtesy of Gary Hudson.

20. After a failed attempt to move the Roton ATV to the Classics Rotor Museum in California, it is now on permanent display at the Mojave Air and Spaceport Rocket Garden. Courtesy of Eric Dahlstrom.

21. Jeff Greason, CEO and founder of XCOR Aerospace, standing in front of XCOR's EZ-Rocket vehicle. Courtesy of XCOR Aerospace.

22. Former astronaut and space shuttle commander Rick Searfoss, one of XCOR's test pilots, standing in front of the EZ-Rocket developed by XCOR. Courtesy of XCOR Aerospace.

23. Above: Peter Diamandis (left) and Bob Richards (right) at International Space University Central Campus, Strasbourg, France, September 2002. Courtesy of Bob Richards.

24. Top right: Rocket models of x prize competing teams in front of X Cup concept design by Patt Rawlings. Courtesy of Ryan L. Kobrick.

25. Bottom right: Brian Feeney, president and founder of the Da Vinci Project (an x prize registered team), at the 2005 X Cup in Las Cruces, New Mexico. Courtesy of Brian Feeney.

26. Burt Rutan, chief designer and founder of Scaled Composites, holds a model of SpaceShipOne. Photograph by Mark Greenberg, courtesy of Virgin Galactic.

27. SpaceShipOne attached to White Knight prior to its first suborbital flight over Mojave, California. Courtesy of Scaled Composites.

28. Mission control at Scaled Composites during the first private suborbital flight. Courtesy of Scaled Composites.

29. Mike Melvill riding on top of SpaceShipOne after the first successful suborbital flight. Courtesy of Robert Pearlman, collectSPACE.com.

30. Brian Binnie and Mike Melvill in front of SpaceShipOne.
Photograph by Mike Mills, courtesy of Virgin Galactic.

31. Ansari X PRIZE successful flight celebration. Courtesy of Paiwei Wei.

32. Per Wimmer with models of WhiteKnightTwo and SpaceShipTwo in front of the Museum of Natural History in New York City. Courtesy of WimmerSpace.com.

33. Top left: Sir Richard Branson standing beside SpaceShipOne
on its maiden suborbital flight, 21 June 2004. Photograph by Mark
Greenberg, courtesy of Virgin Galactic.

34. Bottom left: Sir Richard Branson (left) with Burt Rutan during
the rollout of WhiteKnightTwo, Mojave, California. Photograph by
Mark Greenberg, courtesy of Virgin Galactic.

35. Above: Elon Musk in front of Falcon 9 engines at the SpaceX
launch site, Space Launch Complex 40, Cape Canaveral Air Force
Station, Florida, 8 January 2009. Courtesy of SpaceX.

36. Above: Spaceport America concept design for a new spaceport being developed for Virgin Galactic in Las Cruces, New Mexico. Spaceport America Conceptual Images URS/Foster + Partners.

37. Top right:The Russian-Ukrainian-American launch team, including Michael Gold, Bigelow Aerospace director, Washington DC office (sitting third from left), posing in front of the Dnepr Space Head Module containing Genesis I before it was transported to the silo. Courtesy of Bigelow Aerospace.

38. Bottom right: An artist's conception of Bigelow Aerospace's first Orbital Space Complex. Courtesy of Bigelow Aerospace.

39. Sir Richard Branson in front of WhiteKnightTwo, named *vms Eve* (in honor of his mother), before its maiden flight, Oshkosh, Wisconsin. Photograph by Mark Greenberg, courtesy of Virgin Galactic.

40. Sir Richard Branson (left) and Burt Rutan with *vms Eve* and *vss Enterprise* in the background on the eve of the rollout, 6 December 2009. Photograph by Mark Greenberg, courtesy of Virgin Galactic.

the American segment, tells me that the world changes far faster than we imagine."

Shuttleworth also helped to change the tone of the space experience. The value of ordinary people sharing their experiences and excitement about space had been acknowledged early on in both the U.S. and Soviet space programs, but neither program fully utilized its power to promote space or its national space program.

The emerging role of these space tourists in sharing their intimate experience of space with the general public continued to grow. During Shuttleworth's live BBC interview, he took questions from children around the world. "What did you have for breakfast?" one wanted to know. (Yogurt, a granola bar, and orange juice.) "Do you have to clean your teeth in space?" asked another. (Yes, but he had to get used to swallowing the toothpaste.) He would later share with the press the excitement of watching the plasma glow outside his window during reentry, and what an inaccurate euphemism it was to use the term "soft landing."

Dennis Tito's notion that he could fill the role once envisioned for NASA's Citizen in Space was proving true for another space tourist. Mark Shuttleworth, and each private space traveler to follow, would find a unique way to capture something different from their space experience and share it with the millions who joined vicariously in their adventure.

As the third private citizen to fly with SA, Greg Olsen's mettle and determination were tested above the rest. His dilemma began three weeks into his training at Star City, when doctors found a suspected medical problem on an x-ray of his lungs and disqualified him from his April 2004 flight to the ISS.

"It was a huge blow," Olsen reported to MSNBC. But "that's what builds character."

His medical condition was daunting for a space traveler. He had a history of a collapsed lung, emphysema, and pulmonary tumors. Displaying his usual grim determination, Olsen returned to the United States and spent the better part of a year trying to remedy his condition and convince the Russians that he was medically qualified to train and fly.

With a singleness of purpose and the expertise of an aerospace specialist in Houston, Olsen set about obtaining treatment and rehabilitation.

Therapy included analog environmental tests and simulations. Altitude chamber tests, high-altitude mixed-gas simulation, zero-g flight, and high-g centrifuge runs subjected him to the conditions and environments he would experience on a spaceflight. He also underwent preventive surgery to avoid any occurrence of a collapsed lung.

About a year after being disqualified, Olsen got an urgent call from Anderson telling him to catch a plane to Moscow as soon as possible for a medical review that would definitively get him back into training. He rushed to Moscow and, after checks and rechecks of the now nonexistent spot in his lung, he passed the test.

Olsen was so eager to get back into training that when the Russians told him, "All right, you pass, you're back in the program," he had to ask, "When can I start?" When they said, "How about Monday?," he stayed in Moscow equipped with only an overnight bag. Two days later he was back in Star City training.

"I'm not an astronaut or a cosmonaut. But on the other hand, space is not for wimps," Olsen told MSNBC.com. "It's certainly not just writing a check, going off into space. It's a challenging process. It challenges you physically, emotionally, intellectually, and it's a wonderful experience."

Although he might not be an astronaut—a title fiercely coveted by NASA's cadre of career space explorers in light of this new breed of spaceflight traveler—Olsen took exception to the term "space tourist." Having spent more than nine hundred hours preparing for his flight, he thought he had done considerably more than merely sign up for a tour. He preferred instead the designation "space traveler."

Nor did Olsen feel that the cosmonauts and astronauts with whom he trained resented him as the rich guy buying his way to space. The Russians had a long tradition of selling the third seat on the Soyuz, and he was just another in the line. "The cosmonauts understood that. They saw me in class, they saw me in the gym, busting my tail doing pull-ups and swimming laps. I tried to do the same thing they were doing, and I hoped they respected me for that."

For Olsen, his five months of training felt like a combination of military training and being in college. The "college" part grew out of the camaraderie of sharing the experience with the professionals. The NASA astronauts treated him well, especially Bill McArthur, who took Olsen under his wing.

He shared meals with them and took trips to Moscow together. "I felt privileged," he recalled.

Olsen's preemptive medical conditioning paid off. He experienced no difficulties in his training, performed well during his spaceflight, and suffered no adverse aftereffects. Following the flight, Olsen took the unprecedented step of disclosing his medical records to the space medicine community, a move that gave them valuable information and experience about noncareer astronauts.

Medical disqualification became a critical issue for another SA orbital client, Japanese entrepreneur Daisuke Enomoto. In August 2006, a mere month before his scheduled *Soyuz TMA-9* flight, he was disqualified because of chronic kidney stones. His backup for the flight was an American Iranian businesswoman named Anousheh Ansari. "I was actually going back to my room after finishing training," said Ansari, recalling that moment when she got the good news. "I received a call from Space Adventures telling me that I've been moved up to become part of the primary crew. First I couldn't believe it. I thought they were joking with me."

Enomoto would later file a still pending lawsuit against SA to recover the $21 million he had invested in his training and disputing the reasons for his disqualification. Nevertheless, his removal allowed Ansari to replace him on the September 2006 mission.

Although Ansari had emigrated to the United States from Iran as a teenager, unable to speak a word of English, by the time of her flight she was a successful entrepreneur well known to the space community. Together with her husband, Hamid Ansari, and brother-in-law Amir, the Ansari name became famous when they made a multimillion-dollar pledge in 2004 to the X PRIZE Foundation for its suborbital challenge. The X PRIZE was officially renamed the Ansari X PRIZE in honor of the gift.

Ansari's high name recognition further added to the publicity being generated around the private space industry. As each orbital client flew, it gave them and SA global free publicity through millions of media impressions. At the same time, it reignited the waning spaceflight interest of the general public. Prior to Tito's flight, shuttle and Soyuz flights had become routine. Polls showed that most people didn't even know the ISS existed. But Tito's controversial flight brought space and NASA back to center stage, commanding an unprecedented number of hits on NASA's Web site.

Although most people had been introduced to the concept of private spaceflight, it didn't quite seem real until the fourth orbital client and first female space tourist, Anousheh Ansari, started publishing her online blogs. Previous private space travelers had embraced the mission of sharing space with the public, but no one had done it on this scale, reaching the millions who followed Ansari's diary-like blog during her training and flight. She was an ordinary person being introduced to, and mastering, the concepts of space flight, who was comfortable sharing all her triumphs and anxieties. In fact, she was delivering exactly what NASA had envisioned for its Citizen in Space program. Through her blog, Anousheh Ansari created one of the most intimate portrayals of space flight to date.

Unlike the often mechanical, calculated descriptions from cosmonauts and astronauts, Ansari's blogs gushed with freshness, candor, and emotion. Readers shared the experience of an average person reacting to the incredible experience of spaceflight. The Soyuz launch was "like an airplane take-off"; zero g made her giggle. And her first sight of the whole Earth took her breath away. "Even thinking about it now still brings tears to my eyes. Here it was this beautiful planet turning graciously about itself . . . so peaceful . . . so full of life . . . no signs of war, no signs of borders, no signs of trouble, just pure beauty. How I wished everyone could experience this feeling in their heart."

She shed light on the curious question about what exactly happened in the cramped Soyuz for two days prior to docking. Ansari experienced headaches, back pain, and had to be given a couple of shots for her motion sickness. "It was my fault," she recalled, "I didn't follow instructions. They tell me to avoid very sudden movements and avoid moving too much and try to minimize looking out the window—and I did all that because I was just so excited."

By the time Ansari and her crewmates Michael Lopez-Alegria and Mikhail Tyurin were finally welcomed onboard the station by resident crewmembers Pavel Vinogradov and Jeff Williams (with whom she would later return to Earth), she was all smiles and showed no trace of motion sickness. Of course, readers of her blog learned that it was only because she didn't want to look "like a sick dog" on camera.

Ansari's life onboard the station would be spent performing experiments, conducting video and phone conferences linked back to Earth, and doing simple housekeeping chores. However, her most treasured moments, as reported on her blog, came after scheduled lights out, when the crew finally had time to relax. "I love to just put on my iPod and listen to my favorite music while watching the Earth go by," she wrote. "The station makes a complete orbit every 90 minutes. The sun rises and sets during each orbit, and you can watch 32 beautiful sunrises and sunsets over the course of the day."

On most nights she watched thunderstorms "like a magnificent light show." She then added, "The other night as I was watching this, I was listening to 'Canon' by Johann Pachelbel, and it looked like someone was orchestrating the lightning with the music. But that is not the best part. The best part is the view of the universe at night. The stars up here are unbelievable. It looks like someone has spread diamond dust over a black velvet blanket." She would normally rest her head against the window until it ached from the coldness of the glass.

Even her descriptions of the Station tended to focus on the views of Earth from each ISS window. From the windows in the Service Module, "You can see straight down, so you see just the Earth surface with a little curvature at the edges. From the side windows in the little cabins and the docking compartment, where I sleep, you see the complete curvature of the Earth against the dark background of the universe. This view is actually my favorite because you see the 'Whole' not the 'Parts.' I always like to see the big picture before deciding or worrying about the pieces. I wish the leaders of different nations could do the same and have a world vision first, before a specific vision for their country."

At the end of her ten-day mission, Ansari was reluctant to leave. During their last video conference prior to undocking, her crewmates jokingly said she was hiding in the station in the hopes of staying behind. She said she had butterflies in her stomach and had the same feeling as when she left Iran. But once the hatch was closed, everything was refocused on the landing ahead.

Ansari described the *Soyuz TMA-8* spacecraft entering the atmosphere "like a shooting star." The heat shield threw off an orange glow as the forces peaked at 4 g. "Wow! My face was being stretched in all directions. I must have looked really funny . . . felt like an elephant was sitting on my chest," she reported. After a few moments of peaceful descent, the parachutes opened in three stages, putting them in a crazy spin each time. "It felt like being on one of those spinning saucers or spinning cabin rides in the amusement parks. You are basically thrown all over the place."

Crewmate Pavel Vinogradov announced their descent starting from three thousand meters until impact. They hit the ground and bounced to one side. "Pavel checked to make sure we were all okay. I said everything is great and thanked him for a great landing. Jeff [Williams] did the same thing and as we were hanging upside down in our seats we stretched our arms out and put our hands together to celebrate a safe landing."

While Charles Simonyi was finishing training for his April 2007 spaceflight, *Forbes* magazine named him to its list of the world's billionaires. Although that made it easier to pay the now $25 million cost of a flight, he also came with other qualifications.

Simonyi has referred to himself as the first nerd in space, an allusion to his accomplishments spearheading the creation of Microsoft Office's flagship applications Word and Excel. But he was also a pilot, with licenses for both jets and helicopters, and more than two thousand hours of flying time under his belt. His interest in flying and space travel extended back to his teenage years in his native Hungary, when he won a trip to Moscow as Hungary's Junior Astronaut. There he met one of the first cosmonauts, Pavel Popovich.

Simonyi was able to reconnect with Popovich when the cosmonaut attended the ceremonies at the end of Simonyi's April 2007 mission. "He was very friendly and acknowledged our meeting. I'm sure he had only a vague recollection of it, but I had his signature and the postcard he sent me. Back then, cosmonauts were like movie stars or rock stars. When you look at the equipment that they used, you know that they were real heroes."

Like Ansari, Simonyi used a content-rich Web site (charlesinspace.com) to chronicle his space experience with photographs and video, blogs, and an "Ask Charles" section, where he answered questions from visitors to the site. In a separate "Kids' Space," children could learn about space and Simonyi's space adventure and earn a "Charles in Space Certificate of Achievement."

By the time of Simonyi's 2007 *Soyuz TMA-10* flight to the ISS, Eric Anderson could already boast that SA had created more astronauts than 98 percent of the nations on Earth. "We've turned space tourism from a fantasy, into an idea, that became an industry," he told the BBC. "It's not the dreamers; it's the doers that have created this new market."

Two years later, in March 2009, Charles Simonyi became the first space flight participant to book a second trip to the ISS.

He returned to a much-changed space station. In the interval, the huge Japanese Kibo lab module had been added, as well as the European Columbus module and a final set of solar arrays. No wonder the cost of booking a visit to the ISS had by then risen to $35 million.

Richard Garriott, who finally flew to the ISS aboard *Soyuz TMA-13* in October 2008 as the sixth private citizen to fly in space, made history by becoming the first second-generation American in space. Coincidentally, he was on the ISS at the same time as cosmonaut Sergei Volkov, who was the first second-generation Russian in space. His father was veteran cosmonaut Aleksandr Volkov, who had flown into space on three separate Soyuz missions.

"This mission to the ISS fulfilled a lifelong dream to experience spaceflight as my father first did thirty-five years ago," Garriott said of the experience. "It's an honor to be the first American to follow a parent into space," Garriott said.

Garriott was particularly inspired by his father's space experience, performing scientific experiments onboard *Skylab*, as well as on the shuttle *Columbia* in 1983. Following in his father's footsteps, he scheduled a full scientific agenda for his mission, and spent many hours with his father in preparation for performing the experiments.

In cooperation with NASA, Garriott performed a series of experiments that examined the physical impact of spaceflight on astronauts, such as observing the reaction of the eyes to low and high pressure in microgravity. Since he was the first person to fly in space who had undergone corrective vision surgery, the results of the experiments could influence the restrictions on astronaut applicants who have undergone the same procedure. His other experiments involved studying the effects of spaceflight on the human immune system and on sleep/wake patterns.

Additionally, Garriott conducted a physics experiment and carried with him the so-called immortality drive, a time capsule including a list of humanity's greatest achievements, digitized human DNA, and personal messages. It remains on the station in the event of a catastrophic calamity wiping out Earth.

"While in space, I had the opportunity to conduct scientific experiments and environmental research, but what was most rewarding was speaking to students," Garriott explained in an SA press release. "Growing up in an astronaut family, I firmly believed that every person could go to space, and now I have. I took this opportunity to inspire them with my adventure and let them know they can achieve their wildest dreams, as well, with hard work and perseverance."

The long periods of training required of spaceflight participants, their involvement with onboard science experiments, and the educational duties they have all undertaken make the once popular sobriquet "space tourist" an increasingly inaccurate term. "I absolutely *abhor* it," Garriott told the political blog site *Daily Kos*. "It pains me it has become the standard nomenclature. There have been civilian astronauts and military astronauts, now there are also private astronauts. I can make a good argument for why 'private astronaut' really does apply."

His own flight was not just "an expensive lark," he insisted. "It was an investment in a new and exciting field of private enterprise with the potential to transform our world, one I am completely committed to and lucky enough to be intimately involved with."

When asked if he would do it again, he responded, "Oh, I am going to do it again. Developing space exploration and commercial applications is what I do; it's my career. I and several other high technology business people are investing big in commercial space. We're designing a new generation of reusable rockets, habs [space habitats], and planning the first permanent settlements in space and on the moon or Mars.

"I'm bullish on space. The first space industry billionaire is alive and might even be in their 40s. I'm betting on it, because I'm bullish on us, on human ingenuity. I made several million dollars off my thirty million dollar flight.

"If we can get that cost down to the single digit millions, that's where it starts to become profitable for private companies and research organizations to buy space time. And given our collective thirst for adventure, exploration, and technology, I think that's a bet we're all going to collect on a lot sooner, and far more handsomely, than most people dare to imagine today."

6. The Quest for a Reusable Spaceship

Only free enterprise has the motivation to save money,
and thus only private effort can create a cost-effective
space transportation machine.

Gary Hudson

The winds blew fifty miles an hour in the desert of Mojave, California, on 1 March 1999, buffeting the thousand-strong crowd gathered at the airport. They had been drawn to this isolated Mecca of New Age space development to witness Gary Hudson, CEO of Rotary Rocket, Inc., roll out his Atmospheric Test Vehicle (ATV). When operational, the ATV would be able to deliver access to space faster and cheaper than the space shuttle. In the parlance of the space industry, it would achieve the dream capabilities of vertical take-off and vertical landing (VTOVL), single-stage-to-orbit (SSTO), and full reusability. And it was privately funded. All of that charged the debut with the portent of a bright new future.

"Welcome to the Revolution!" yelled master of ceremonies Rick Tumlinson, cofounder of the Space Frontier Foundation and of MirCorp. "The opening of space is about dreams," he told the crowd, "the dream that was written about by Heinlein, Clarke, O'Neill, and others."

Other notable speakers also marked the significance of the occasion. "This is America," charged Tom Clancy, author and Rotary Rocket benefactor. "We are the guys who change the world. We are the guys who don't know or don't care what 'impossible' means." Federal Aviation Administration (FAA) associate administrator Patti Grace Smith explained how government regulations were evolving to support the development of innovative launch vehicles. NASA chief engineer Daniel Mulville drew a connection between NASA's mission to explore space and the mission of private industry to exploit it.

Then, to the chords of Aaron Copland's *Fanfare for the Common Man*, the hangar door slowly opened. High winds blew away the expected dramatic effect of the fog machine as Rotary Rocket's staff rolled out of the hanger the innovative spacecraft known as Roton.

After the formalities and photographs, the crowd moved in for a closer look. The white, conical-shaped ATV stood sixty-four feet high and twenty-two feet wide, with a single huge side window near the base. Scaled Composites, the company of aviation visionary Burt Rutan, had been contracted to build the strange craft. It resembled a number of SSTO projects that had come before it, but with one notable difference — it sported helicopter-type propeller blades atop the nose cone.

It was estimated that when operational, the spacecraft would be capable of launching seven thousand pounds of payload into low Earth orbit for $1,000 per pound. This would be quite a launch bargain for the booming wireless telecommunications satellite market. That lift capability would also allow it to carry a dozen people, making it wonderfully suited as well for the space tourism market.

The project had its genesis about three years earlier, when Hudson sat in a conference room at American Rocket Company, a firm cofounded by his friend Bevin McKinney. McKinney was enthusiastic about this wild idea to build a space helicopter — a rocket ship powered by a huge propeller. Hudson would later write about that pivotal moment in a *Wired* magazine article. His first reaction was to state the blunt truth: "Bevin, that's insane." His second reaction was to keep listening. Hudson had been in the space business long enough to know that "the difference between 'insane' and 'insanely great' is often only a matter of shifting perception."

Motivating both men to consider crazy ideas was a desire to end the government monopoly on access to space. "I had been frustrated that space was the exclusive province of government space agencies and hero astronauts. We both wanted to go to space for the sheer fun of it and, in the finest capitalist tradition, make a few bucks along the way," Hudson explained.

At that point in time, bucks were the one thing that neither man was making. According to Hudson, both of their companies were being driven out of business by government-funded competitors. The only way to tackle the problem, Hudson concluded, was to build a single-stage reusable rocket. It would be easy to maintain, reliable, have a fast turnaround time, and, most important, would be accomplished without any government funding. He estimated that he would need about $150 million for three prototypes, licensing, and testing.

His main financial backing came from two men who shared Hudson's desire to end NASA's monopoly on space. Walt Anderson, president of the business development company Gold & Appel, had already given generously to other space projects. In 2000 he would establish the Foundation for the Non-governmental Development of Space. He learned about Hudson's project from the *Wired* magazine article and personally sought him out. After three meetings with Hudson, Anderson decided to provide financing for the project, maintaining that it was the easiest investment decision he'd ever made.

The other benefactor, Tom Clancy, who initially invested $1 million, had his own reasons for getting involved. Known for his conservative, sometimes contrarian views, he had come to despair of NASA's lack of vision. Roton was the alternative, an example of the private, entrepreneurial, can-do spirit that would deliver inexpensive access to space without big, wasteful government. He would be happy if Hudson put NASA out of business.

Justifying their faith in him, Hudson had taken McKinney's seemingly crazy idea from concept to actual hardware rollout in the short span of three years.

Getting hooked on space was a common affliction for those growing up in the 1950s, and Gary Hudson was no exception. Born in St. Paul, Minnesota, in 1950, he indulged in space books and had a steady diet of science and science fiction programs like Disney's *Man in Space* TV series during the Apollo era. Hudson was comforted by the fact that at the beginning of the space age, no kid would be labeled a dreamer for expressing an interest in space.

Hudson read Arthur C. Clarke's visionary book *Profiles of the Future* when he was ten. "In that optimistic look into the idea of the 'limits of the possible,' Arthur ignited my desire to be at the forefront of the opening of the space frontier." Hudson admits that if he had realized back then that fifty years later we would still be struggling to achieve those dreams, "I likely would have been too discouraged to begin."

However, it was the reading of another book, in 1969, that fired Hudson with the passion for private space. The book was Philip Bono's *Frontiers of Space*, coauthored with Kenneth Gatland. Bono was considered the father of the SSTO craft. An aeronautical engineer who had analyzed captured V-2 rockets, he worked for Boeing for nine years on the Air Force's X-20 Dyna-Soar project. The X-20 was a reusable, one-man, rocket-boosted orbital glider, a precursor to NASA's space shuttle. Unfortunately, the massive development costs and questionable usefulness of the military spaceplane led to the project's cancelation in December 1963.

Bono later became the technical director for Douglas Aircraft during NASA's Apollo and Voyager programs in the 1960s. As director of launch vehicle studies, he was the first to propose recovering and reusing all three stages of the Saturn V rocket. In fact, he argued that the third stage of the Saturn IV-B had sufficient thrust as an SSTO to send an eight-thousand-pound Gemini capsule into orbit. He called it a "Saturn Application Single Stage to Orbit," or SASSTO. The SASSTO would be the precursor to the Delta Clipper and the DC-X programs of the early 1990s.

Throughout the 1960s, Bono worked on numerous SSTO designs, convinced that these craft were the solution to cheaper access to space. His projects included the Reusable Orbital Modular Booster and Utility Shuttle (ROMBUS), an intercontinental passenger vehicle called Pegasus, a troop-carrying rocket called Icarus, and a rocket sled–propelled design known as the Hyperion SSTO.

To Hudson's way of thinking, Bono had got hung up on the "bigger is better" concept of designing spacecraft, and consequently kept running into technical design problems to satisfy the weight requirements. Hudson, on the other hand, had come to the conclusion that the "simple SSTO" was the solution to cheaper access to space.

After reading Bono's book in 1969, Hudson recalls, "I immediately wrote Phil a letter, saying that while I was impressed with his concepts, he was missing the boat by thinking the government would fund these new ideas." Instead, Hudson argued strongly that such a craft should be privately funded, and he offered to find the funding.

The idea of private funding had only recently been broached in an American Astronautical Society conference on the commercial use of outer space. That and the timely release of *2001: A Space Odyssey* with evocative images of Pan Am spaceliners and trips to the moon had fueled Hudson's ideas of space commercialization. On several occasions he visited Bono in Costa Mesa, California, to discuss Bono's technical ideas and vehicle concepts, while continuing to urge him to go after private financing for SSTO development. "Bono was discouraged by the state of the aerospace industry in those years as Apollo funding collapsed," Hudson recalls, "and so was unconvinced that private dollars could be found."

In 1972 Hudson visited London to present his case for private funding to Bono's coauthor, Kenneth Gatland, who at the time was executive director of the venerable British Interplanetary Society. It was another bold move, given that Hudson had little to offer in the way of credentials. He was a mere twenty-two years old, and one year earlier had dropped out of the University of Minnesota to pursue his dream of building rockets. Therefore, unlike most spacecraft designers of his time, he did not even have an engineering degree.

Gatland met with Hudson and patiently listened to his ideas regarding private SSTO development. To set Hudson straight about the difficulty of what he was proposing, Gatland sent him to talk to Val Cleaver, then chief engineer of rocket engine and power plant development for Rolls-Royce. At a subsequent lunch meeting in London's Piccadilly Circus, Hudson remembers Cleaver saying, "You can no more build an SSTO than you could levitate yourself over Piccadilly Circus this instant." He then added, "And I have a friend in town who'll tell you the same thing."

The friend turned out to be none other than Arthur C. Clark, home in London on family business. A few days later Hudson met Clarke at his club. "After he heard me out, he gently disagreed with his friend Cleaver and remarked that he thought I had a chance, if a small one, to achieve the goal of building a reusable space transportation system. The key was to be willing to try."

After returning from London, Hudson tried without success to convince Bono to assemble a team and develop a business plan. "Unfortunately, his career had reached a dead end," Hudson recalled. "The fight had gone out of him, and I can fully appreciate what he must have been feeling then, since I have experienced it myself. But I resolved to carry on, giving him credit for inspiring me, but also determined to improve on his concepts, taking advantage of new technology and ideas I have developed over the years since I first read his book."

By May 1974 Hudson had founded his first company, Space Merchants, Inc. He is actually considered one of the "first space entrepreneurs," as he was one of the first to realize that space travel equates to profitability. "I tried to consider all of the things about space launch that added cost," Hudson noted. "It seemed obvious that both expendability and staging were incompatible with low-cost operations. The oft-quoted analogy to throwing away a commercial aircraft after one use is apt, if trite." He then adds, "I never saw the government develop a commercial air transport. Only free enterprise has the motivation to save money (rather than create jobs) and thus only private effort can create a cost-effective space transportation machine."

Because of such sentiments, Hudson was considered an outsider by many of those within the rocket community. The purists among them did not buy into his grand scheme of commercializing space. Government monopoly and government funding for space projects was still the only game in town.

Hudson's first designs ran contrary to the norm. His company designed the Osiris, a vtovl spacecraft capable of lifting twenty-four thousand pounds of payload into orbit. What was unique about the design at that time was the placement of the payload compartment between the liquid hydrogen tank in the nose and the liquid oxygen tank below. It was an innovative departure from the standard artillery shell rocket design and the first time this configuration was ever used. Later it would become the precursor for other ssto rocket designs.

Hudson's subsequent designs proved that big isn't always better. His Phoenix craft was designed to carry 10,000 pounds to orbit, and its scaled-down version, the atv, with a payload of 1,100 pounds, was specifically designed to cater to the emerging satellite market. But in the 1970s, no one would fund the seemingly radical designs Hudson proposed. "The 1970s were tough years," he concedes. "Of course, everyone expected the Shuttle to live up to its promised reduction in launch costs. Most people I spoke with didn't think a commercial launch system of any type was needed."

By 1980, under the company name gch, Inc., Hudson finally was contracted to build a launch vehicle by Space Services, Inc. of America, a company owned by Houston real estate magnate David Hannah Jr. Hannah wanted a cheap, expendable satellite launcher to serve as an alternative to NASA's booked-up space shuttle for launching telecommunications and earth-scanning satellites.

The rocket he created, the Percheron, was considered the first privately funded space launcher project in the United States. Principal to its design was a simple, pressure-fed, kerosene-oxidizer engine that was intended to reduce the massive costs associated with discarding the booster rocket.

The project ran into export control problems in 1981. Hudson had planned to launch the Percheron from Matagorda Island, a barrier island situated on the Texas Gulf coast, which would mean launching into international waters of the Gulf of Mexico. Houston lawyer Art Dula helped Hudson obtain the permit and licensing to fly the Percheron, but as it turned out, they would not need them. On 5 August 1981 the Percheron blew up on the launchpad during a static firing as a result of a component malfunction. It was a bitter pill for Hudson to swallow and was seen by some as vindication for NASA and the professional rocket community, who never believed the rocket would ever get off the ground.

David Hannah subsequently pursued the project with Space Services, Inc., a company comprised of former NASA engineers, as well as Mercury astronaut Deke Slayton, who had recently retired from the space agency. They would come up with an entirely new booster design, which was based on clustering engines from the second stage of the Minuteman missile. On 9 September 1982 their rocket, *Conestoga 1*, launched from the missile range located at Wallops Island, Virginia. The rocket consisted of the core missile stage and a 1,100-pound test payload. The payload was successfully ejected at an altitude just short of two hundred miles. On that date, therefore, *Conestoga 1* effectively became the first privately funded rocket to reach space.

Meanwhile, Hudson went back to work, reviving the Phoenix SSTO project and once again working with a negative budget. It was 1984, the heyday of the space shuttle, and President Ronald Reagan had just announced that NASA would fly an ordinary citizen, a teacher, into space.

Around this time, T. C. Swartz, operator of Society Expeditions in Seattle, Washington, hit on the bold, imaginative idea that flying people to space would become the next ultimate travel adventure. He approached Hudson to build him a passenger-rated orbital spaceship for his Project Space Voyages. Hudson accepted the challenge and immediately hired one of his mentors, aerospace engineer Maxwell Hunter, who was then working for Lockheed Martin as a consultant. The challenge would be to redesign the Phoenix to suit Society Expedition's requirements.

Hunter brought to the project over forty years of aircraft, missile, and space experience working for Douglas Aircraft Company and Lockheed. He was responsible for designing and launching the Thor missile, which later evolved into the McDonnell Douglas Delta rocket, the most widely used expendable rocket in history. He was also involved with a nuclear-propelled ssto rocket called Rita until 1969, and with the Lockheed ssto X-Rocket in 1985.

Like Hudson, Hunter had his eye on sstos to improve space transportation. Both were members of the Citizens' Advisory Council on National Space Policy and had campaigned heavily for the concept of sstos. The council was founded in 1980 to influence the new Reagan administration policies on the future of the U.S. space program. Members included such visionaries as the noted science fiction writers Robert Heinlein, Larry Niven, and Jerry Pournelle.

Back at GCH, Hudson realized that the market for adventure travel to space would probably not be large enough economically for Swartz to buy and operate his own fleet of vehicles. Instead, he agreed to lease his Phoenix sstos and use them for other markets himself. Both Hudson and Swartz tried hard to find investors to develop and build the vehicle, but to no avail. "I used to say it was because every investor had a brother-in-law in the aerospace business," Hudson recalled. For each investor trying to do "due diligence" on the untried and new technology, there would always be NASA experts available to give their opinion. In those days the standard answer would always be the same: "It won't work. If it would work, NASA would be doing it." The project never got off the ground, and the loss of the shuttle *Challenger* in January 1986 sealed the project's fate. "I never got even a nickel from Society Expeditions," Hudson mused.

For Hudson, however, pursuing the dream of building a reusable launch vehicle (RLV) was not lost. In 1988 Maxwell Hunter proposed to the Citizens' Advisory Council on National Space Policy an ssto concept called SpaceShip Experimental (ssx). The ssx was reminiscent in many ways of Phillip Bono's spaceship design concepts, Hudson's Phoenix ssto, and his own X-rocket. It was designed to be built in three years for less than $1 billion, carry a ten-ton payload into low Earth orbit, and have a twenty-four-hour turn-around time.

In his book *Halfway to Anywhere*, Citizens' Council member G. Harry Stine recalled one appealing aspect of Hunter's ambitious plan. "It would be produced for commercial sale to the government and to any corporate entity that wanted to build and fly spaceships." More to the point, for those who might be interested in private space development, the ssx promised to be fully reusable, reliable, and inexpensive—everything the shuttle was not.

On 15 February 1989 the council delivered Hudson's concept and white paper for the ssx to the Executive Office Building of the White House. Hunter and Pournelle, along with Daniel O. Graham, founder and head of High Frontier, an organization advocating for a space-based missile defense system, met with and received a positive response from Vice President Dan Quayle, then chairman of the National Space Council. Their white paper made the valid argument that any U.S. space-based weapons system would need to be serviced by a reusable ssto spacecraft with a higher reliability factor than the then grounded shuttle fleet, in addition to offering lower launch costs and consistently better turnaround times. To their relief, the vice president concurred with their findings and set things in motion.

It would prove to be the beginning of a new era for developing simple and efficient rlvs for the government as a precursor to the commercial industry that was to follow.

As a result of the meeting with the vice president, an ssto program was initiated and spearheaded by the Strategic Defense Initiative Organization (sdio) office of the U.S. Air Force under its director, Lt. General George L. Monahan Jr. Monahan assigned his deputy for technology, Lt. Colonel Simon P. "Pete" Worden, to set up the ssto office. The task of creating the program fell to two officers, Lt. Colonel Pat Ladner and Major Jess M. Sponable.

The program got under way with a $15 million concept study on existing ssto technology. A subscale suborbital test vehicle would then be built to prove the technology, followed by a full-scale experimental vehicle capable of launching to orbit. Unlike the billions of dollars that sustained some nasa projects, this would be a bargain-basement program. Whoever won the contract on the subscale test vehicle would have a comparatively modest $60 million to design, test, and fly this prototype of a fully reusable RLV that could lift a twenty-thousand-pound payload into orbit. As a consequence, this would leave zero dollars for research and development, which meant that the vehicle would have to use existing technology.

There were four final competitors for the study competition. The first was General Dynamics's VTOVL rocket, the Millennium Express, with the subsequent experimental vehicle called the Pathfinder. Then there was McDonnell Douglas's VTOVL rocket, the Delta Clipper Experimental (DC-X) and experimental vehicle DC-Y. Another competitor was Rockwell International's Vertical Take-Off Horizontal Landing (VTOHL) spaceship, based on the space shuttle orbiter technology; and finally Boeing's Horizontal Take-Off Horizontal Landing (HTOHL) concept that could be air-launched from a carrier or ground-launched by a rocket sled.

McDonnell Douglas ultimately won a $58.9 million contract in August 1991 to develop its rocket design within two years and construct a scale prototype. In close consultation with the Citizens' Council, the one-third scale demonstrator vehicle DC-X was to be built under the leadership of William A. Gaubatz, director of SSTO programs at McDonnell Douglas. The company would develop its rocket under the project name of Delta Clipper, or DC-X, in honor of the company's successful Delta rocket.

The Delta Clipper closely resembled the conical-shaped SSTOs previously designed by Hunter and Hudson, apart from possessing more streamlined, flattened sides. However, the DC-X was never designed to achieve orbital altitudes or velocity. Instead, its development was focused on actually demonstrating the concept of vertical takeoff and landing, as popularly portrayed in science fiction movies of that era. The reusable DC-X was developed to take off vertically, but in contrast to conventional rockets, it would then return to the ground in a controlled vertical landing, using attitude control thrusters and retrorockets during the descent, before settling down on base-mounted, retractable landing struts. The Delta Clipper could then be refueled and made ready once again for liftoff.

Burt Rutan's Scaled Composites was contracted to build the DC-X's structural aeroshell and aerodynamic control surfaces. At sixty-two feet in height and a maximum fifteen feet at the base, the DC-X would be powered to twenty thousand feet by four Pratt and Whitney RL-10 A5 rocket engines, each generating 13,500 pounds of thrust.

Things seldom run smoothly in the big-project rocket business, where funding and political will must extend over many years. By 1992 the DC-X had its problems, as the project's advocates began falling by the wayside. Dan Quayle's failed reelection bid came as a blow to the SSTO project, while the Pentagon/NASA advocates of expendable launch vehicle development had started to tip the balance of power by downplaying the potential success of the DC-X and doubting its feasibility.

The SDIO office found itself renamed the Ballistic Missile Defense Organization (BDIO), while Mike Griffin, its chief engineer and a strong SSTO advocate, was reassigned to head NASA's Office of Exploration, a post with a title but no portfolio. Furthermore, the limited funds for the project had forced Pete Worden to cancel development of the orbital DC-Y and reallocate funds to finish and test one DC-X prototype. Despite a general feeling that the SSTO program would literally never get off the ground, the DC-X project would prove the doubters wrong twenty months after its conception.

McDonnell Douglas's project manager for the DC-X Delta Clipper, William Gaubatz, told the space magazine *Interavia Space Markets* during the rocket's development stage that it was too early to say whether the reusable booster could compete financially with existing but expendable, solid-fueled sounding rockets. However, he added, "If you look through some of the brochures of current sub-orbital rockets, they are quite a totem pole of combinations of surplus solid motor stages. They're expensive, and you are at the mercy of the quality of the surplus stager that is given to you."

Furthermore, Gaubatz insisted that "it is in the interests of the government to help establish the United States as a very strong competitor in the world market." In analyzing his comments, *Interavia Space Markets* commented, "It should also be in the interests of McDonnell Douglas."

With the space shuttle now clearly out of favor with the White House, it was conceivable that President George H. W. Bush might even order NASA to take on the responsibility and task of building an orbital version of the Delta Clipper as a replacement manned vehicle. However, NASA doggedly dug in its heels, completely reluctant to take on a program that had originated elsewhere, especially at a time when savage budgetary cutbacks were eating into its own programs. It was a time of high tension, eventually resulting in NASA administrator, former shuttle commander Richard Truly, handing his resignation to President Bush on 10 February 1992, reportedly under pressure over these issues.

A year later, on 3 April 1993, to the unbridled jubilation of its development team and the applause of VIP guests, a gleaming forty-two-foot, one-third-scale Delta Clipper DC-X test vehicle was majestically rolled out at McDonnell Douglas's Space Systems Facility in Huntington Beach, California. To that point the project had consumed only $50 million in development costs. Dressed in red overalls, the twenty McDonnell Douglas ground crew members served to remind those in attendance of the stark difference between what it takes to run a small launch team operation, compared to the thousands of personnel needed to launch a space shuttle.

A few weeks after the rollout ceremony, the evolutionary DC-X test vehicle, a precursor to the proposed DC-Y, was transported to White Sands Test Facility in New Mexico, where it was hot-fired during two months of static tests. For its launch and landing pads, the DC-X would use two concrete pads originally constructed to support the cranes that lifted the space shuttle atop a Boeing 747 carrier aircraft.

On the afternoon of 18 August 1993, the first flight of the DC-X took place in anonymity, witnessed only by DC-X staff and crew. The obelisk-shaped vehicle was piloted remotely from the flight operations control area by a crew of three, headed by former Apollo astronaut and McDonnell Douglas vice president Charles "Pete" Conrad Jr. By comparison, a space shuttle launch employed a team of around 1,700.

The vehicle rose to an altitude of 150 feet above a plume of yellow flame and hovered in mid-air. It then flew sideways 350 feet while maintaining altitude, stopped, then descended and deployed its landing gear at about 100 feet, touching down on the second concrete pad. Conrad yelled, "Touchdown! Touchdown! Weight on gear! Weight on gear! Engine shutdown."

The second Delta Clipper flight took place three weeks later, on 10 September. This time there was an audience of VIP guests and news media, seated in bleachers three and a half miles from the launch pad.

As Caleb John Clark wrote for the *Los Alamos Monitor* the following day, "Rising 300 plus feet straight up, the DC-X did something rockets aren't supposed to do: it stopped in mid-climb, and hovered, absolutely still, for three seconds. It then pitched almost imperceptibly to the left and traveled horizontally for 300-plus feet. The DC-X then stopped and hovered again and started its descent. At about 200 feet above its landing pad four long landing struts appeared out of its base and as the vehicle neared the Earth, yellow flames flared as it came to a soft landing, 66 seconds after its take off."

Flight manager Peter Conrad was visibly exuberant, cracking a wide, gap-toothed smile when asked by a reporter what event in history this might parallel. "We landed vertically on the moon," the *Apollo 12* moonwalker stated, "and we're landing vertically now, which with the [Earth's] atmosphere is tougher."

Despite the mounting opposition from SSTO opponents, the DC-X team would conduct eight hover test flights until 1995, reaching a maximum height of 2,500 meters. The craft proved that it could maneuver in flight, in any direction, and be ready to fly again within twenty-four hours.

Even the traditional space community found these results hard to ignore. In 1994 President Clinton's U.S. Space Transportation Policy transferred control of the project to NASA, which seemed to bode well for its future. "Delta Clipper is one of the most imaginative programs that this nation came up with for access to space," NASA administrator Dan Goldin told the National Press Club. "We are enthusiastic about it."

In 1996 Goldin even accorded General Daniel Graham NASA's highest award for his early role in championing the SSTO concept, and renamed the DC-X the Clipper Graham. For once, it seemed that NASA and the private space community were on the same page, both supporting DC-X and Goldin's organizational mantra of a "faster, better, cheaper" approach to space.

But then occurred one of those developments that fuel conspiracy theories among private space advocates, that NASA is a cabal of entrenched interests less concerned about space than about maintaining programs and jobs.

Two years earlier, in 1994, and seemingly working against the express interests of its administrator, the Marshal Space Flight Center had issued a request for proposals for the X-33, a vehicle to replace the shuttle.

The X-33 would be a subscale technology demonstrator, larger and supposedly less expensive to fly than the Delta Clipper. Lockheed Martin won the contract, announcing it would develop a craft called the Venture Star — said to be the next generation of space vehicles. Suddenly, the DC-X had been robbed of its future.

On the last flight of its second test series in 1996, the vehicle, now called DC-XA, suffered heavy damage when part of the landing gear buckled on touchdown. The vehicle was one-of-a-kind and thus irreplaceable. Advocates in the private space community lobbied hard for continued funding, but it was not to be.

Meanwhile, Lockheed Martin spent five years and burned through almost $1.5 billion in taxpayer dollars, and never flew a single piece of hardware. The hopes and dreams of the SSTO community were again at a standstill.

On 11 October 2001 Dr. Henry F. Cooper, chairman of the High Frontier organization and civilian director of the Strategic Defense Initiative from 1990–93, shared his views on the need for a viable science and technology program to build spaceplanes in a passionate speech he gave before the Space and Aeronautics Subcommittee of the House Science Committee. In his testimony he reviewed the history of the DC-X program and recalled how General Graham had originally taken the SSTO concept to Vice President Quayle, "who in turn, requested that the SDI Organization undertake a serious development program. They chose SDIO because, as General Graham wrote in his memoirs, they expected that either the Air Force or NASA 'would stifle our baby in the crib' — NASA because it threatened shuttle prolongation and the Air Force because it threatened its next expendable rocket program.

"They were right then," Gordon stated, "and I suspect their view of both NASA and the institutional Air Force is still correct."

By the time Gary Hudson's company Rotary Rocket was in operation in 1996, he had already founded and folded six different rocket companies. This time, he would need to raise about $150 million to fund a full-scale reusable Roton rocket. Walt Anderson, his primary benefactor, had persuaded Hudson not to enter the Roton in the X PRIZE competition because of personal animosity between Anderson and founder Peter Diamandis.

The X PRIZE was a competition to produce the first nongovernment, reusable manned spacecraft and successfully launch it into space twice within two weeks. It carried a $10 million prize. Roton had originally been designed to be an X PRIZE competitor. Turning away from that goal required a dramatic change in the project's focus. Although still advertised as one of the potential vehicles that could be used by Space Adventures on their suborbital flights, Hudson knew he would have to go after the low Earth orbit communication satellite market in order to attract additional investors. Hudson approached venture capital firms, investment banks, and potential customers, but with the collapse of the satellite market, he had little success raising money.

The Roton ATV managed to fly three times in 1999. Hudson hired two former naval test aviators, Brian Binnie and Marti Sarigul-Klijn—respectively the company's flight test director and chief engineer—to fly his rocket. On the date of the first flight, 23 July, Binnie was worried. "I guess I'm about as scared as I was on my first carrier landing. More scared than I was the first time I jumped out of a plane."

However, the first flight was successfully completed by copilots Sarigul-Klijn and Binnie in a crucial test to assess the feasibility of the final stage of the landing. The four propeller blades lifted the Roton rocket to a height of eight feet, where it hovered for a few seconds and then was brought back down to a gentle landing. "The last six inches are the exciting ones," spokesman Geoffrey Hughes told a cluster of reporters. "We've demonstrated that we can land this thing."

On its second flight, on 16 September, the spacecraft flew a 2.5-minute flight to a maximum height of 20 feet. On its third and final flight, on 12 October, the Roton flew down the flight line at Mojave Airport, reaching a height of 75 feet and traveling 4,300 feet downrange, at speeds as high as 53 mph.

Hudson recalled thinking that Sarigul-Klijn was about to lose control on the final flight. "It was a devil of a beast to fly. (Of course, it wasn't really meant to be flown that way on a return from orbit.) He had taken a few years off the end of my life by hovering for many minutes after the main part of the test was concluded."

To the civilian space community, seeing an actual vehicle being tested and performing quite well reignited a glimmer of hope and excitement that cheap access to space was indeed around the corner. It inspired other space entrepreneurs and vehicle developers at that time to soldier on. On the other hand, there was considerable criticism over the design of the vehicle.

Visibility in the cockpit was restricted, which made flying it extremely difficult. The two pilots likened the cockpit to a "bat cave" and had to rely on a sonar altimeter to determine their altitude because of their obstructed view of the ground. Furthermore, because of the torque generated by the spinning rotor blades, the spacecraft had a tendency to spin, which had to be counteracted manually with thrust applied in the opposite direction.

Difficulties were apparent even flying the simulators. On a scale of one to ten (with one being simple to fly and ten being so difficult to fly that a crash is virtually imminent), more than seventy civilian and military pilots who tried the simulator rated flying the Roton a ten. There was much speculation about whether a fully developed Roton could actually work.

Hudson's accelerated rocket development project included the daunting task of building a new rocket engine, dubbed "RocketJet." He hired a number of young engineers, such as Brent Eubanks and Christopher Smith, to work on the problem under the leadership of engineer Jeff Greason. The plan after the ATV was to build the Propulsion Test Vehicle-1 and Vehicle-2, which would first demonstrate full rotor system capability to a height of five hundred thousand feet and eventually test suborbital and orbital capabilities.

But financing was always a problem. By the end of 1998, Hudson had raised only a fraction of what he needed to develop a vehicle. Capital was bleeding fast into overhead and Scaled Composites's development. Even before planning commenced on the ATV rollout, Hudson knew there was only enough money to debut and flight test the ATV, not enough to develop the full-scale RLV. Hudson was banking on the media exposure and publicity of the rollout to lure new investors to the project. But that never happened. Even Rotary's main investor, Walt Anderson, who sank another $3 million into the project, grumbled as he left the rollout that "nobody will ever invest in Rotary. Only a fool like me would, because it makes no sense."

By 1999 the huge market for small communication satellites and the ensuing need for launching them had collapsed. As former Rotary Rocket engineer Tom Brosz explained years later, "Our target niche vanished almost overnight, and at this stage of the game, nobody was looking at space tourism, something the small Roton was almost made for." Attempts were made to redesign the vehicle for a bigger payload, but that presented too many technical problems and design fixes that were well beyond the support of the initial business plan.

"Attempts to raise money from more investors did not go well," Brosz said. "At the time, we were competing with a healthy dot.com industry, and why put cash into something that will make you rich in five years, when you can put your cash into something that will make you rich in a month? That, and the amounts required had grown with the vehicle size. Now, for all intents and purposes, we were now building the DC-Y [the next stage of the DC-X], and we didn't have those kind of resources."

Rotary canceled further testing and ceased engine development in 2000. Hudson had raised about $33 million but had failed to secure launch contracts. As a result, Rotary Rocket folded in 2001.

In retrospect, Hudson could cite many reasons for Rotary's failure. "We tried something too ambitious. We didn't have all the funds to reach a flight goal and marketability in hand when we began. I couldn't hire sufficiently experienced people since I couldn't promise them jobs even six months into the future. And I had to spend my time looking for money and managing the overall company and the vehicle engineering. Each of these three tasks is more than a full-time job and obviously can't be done by one person."

After a failed attempt to transport the ATV to the Classic Rotors Helicopter Museum in San Diego, California, it was moved to its permanent home at Mojave Airport in November 2006. Today, the Rotary hangars are occupied by the National Test Pilot School.

To the outside world, the story of Rotary Rocket was yet another addition to the list of failed reusable rocket developments over the years. But to Hudson and the people who worked on the project, flight testing the Roton had proved that a small company with limited funds could get to this stage of vehicle development without government support—a far better accomplishment than such multibillion-dollar projects as Lockheed Martin's failed X-33 during the same time period. Seen in that perspective, Roton's achievements helped the private sector to stake its claim on the space age.

Today, Hudson is involved in other people's start-up ventures but does not completely discount the possibility of one day returning to develop an SSTO. But only if he has all the money before he starts.

As Hudson was reluctantly laying off people at Rotary Rocket in 1999, some of his development team decided to take their start-up experience and create a brand-new space company. Thus, xcor Aerospace was born just as Rotary Rocket was on the verge of closing up shop.

The four xcor founders were Rotary's first employee, Aleta Jackson, and three of its engineers—Dan DeLong, Doug Jones, and Jeff Greason. De-Long and Jones would serve as chief engineer and test engineer, respectively, while Greason and Jackson would take on the task of managing the company as president and manager. They rented one of the hangars at Mojave not far from Rotary Rocket and Scaled Composites and started operations using their personal savings.

xcor's company philosophy, like Hudson's, was to make sure that rocket-powered transportation became commercially viable. Undoubtedly learning from Rotary's too-big-too-soon approach, xcor preferred to start small and gain experience, while slowly but surely developing safe, reliable, routinely operated rocket engines for commercial transportation. As Greason explained, "From xcor's beginnings, we have been guided by the vision of rocket operations being as routine as any other form of transportation."

To achieve such a goal, xcor started by developing and producing the NEX-1, an upgraded version of the Bell x-1 rocket plane that Chuck Yeager flew to break the so-called sound barrier for the first time in 1947. xcor would build an updated version of the Bell x-1's engine, the xlr-11, which would help jump-start and fund the company's engine development program. The price tag to own the very first private rocket plane would be $5 million.

Two months after incorporation, xcor had tested and produced a reliable engine ignition system, a critical component for the safety of any vehicle launcher system. They went on to develop a series of small-scale, fully integrated rocket engines, starting with the xr-2p1 "Tea Cart" engine. The engine, named for its installation on a roll-around cart and the heated cooling water it produced, was capable of producing fifteen pounds of thrust using nitrous oxide gas and alcohol.

In April 2000 Greason and his engineering team gave a public demonstration of the Tea Cart engine at the Space Access Society's annual meeting in Scottsdale, Arizona—with full permission from the fire marshal and the Holiday Inn management. The motor was successfully fired several times in the hotel's conference room. At the conference, xcor signed on its first two investors.

By the end of 2000, XCOR had already started work on its regenerative-cooled four-hundred-pound-thrust liquid oxygen (LOX)/alcohol engine called the XR-4A3. The plan was to fully integrate it into an experimental vehicle that could demonstrate a safe and reliable, complete aircraft rocket propulsion system that was both cheap and operable. The airframe of choice was Rutan's homebuilt kit aircraft, the Long-EZ. With a twenty-six-foot wing span and signature Rutan canard configuration, the choice was based on the EZ's pusher configuration (engine mounted forward of the propeller) and good power-off glide capability. Dan DeLong's personal Long-EZ was then modified to accommodate a twin liquid propellant rocket propulsion system. It would be called the EZ-Rocket. Now that they had the engine and the airplane body, the only thing left was to find a test pilot.

"We went looking for a pilot and naturally thought of Dick Rutan," Delong said. Dick Rutan, Burt Rutan's brother, was known as one of the best—if not *the* best—civilian test pilots in the world. He flew the *Voyager* aircraft around the world on one tank of gas with copilot Jeana Yeager in 1986. But would he fly a rocket plane? Delong well remembers when he first pitched the idea to Rutan.

"At first he got real quiet, then said, 'An EZ with a rocket, huh?' Then he said it again, 'An EZ with a rocket, huh!' The next day he called and said, 'Yeah, why not?'"

Prior to the AirVenture Oshkosh 2001 air show, the largest aviation display show in the world, the EZ-Rocket was slated to be put on static display at the event because it had not yet been test flown and only had one engine. But to the surprise of everyone, Dick managed to perform an unscheduled test flight over the Mojave runway just before the show. Later Dick explained, "I'm not going to be associated with a company at Oshkosh that brings an airplane that hasn't flown." Not long after, the EZ was put on a trailer and driven thirty hours nonstop to Oshkosh by three XCOR employees.

Test pilots Mike Melvill and Rick Searfoss would eventually fly the EZ-Rocket in Mojave. It made its official public debut at Oshkosh the following year. Among other firsts, the EZ demonstrated shut-down and restart of the rocket engine in mid-air and a "touch and go" maneuver, which had previously never been attempted or achieved on a rocket-propelled flight. At Oshkosh, Dick Rutan flew the EZ to an altitude of nine thousand feet for ten minutes with a bare-bones ground crew of five people.

The EZ was built with $500,000 in capital over a period of one year and cost approximately $900 per flight. Named by *Time* magazine the "Transportation Invention of the Year," in 2001 the EZ-Rocket successfully demonstrated low-cost and low-maintenance rocket-powered flight. The EZ-Rocket would eventually be flown twenty-six times, including demonstrations at the countdown to the X PRIZE Cup rocket festival event in Las Cruces, New Mexico, in 2005, where it set the world record for distance without landing in its class that year.

A few days after the debut air show demonstration in 2002, XCOR announced a marketing partnership with Space Adventures, Ltd. XCOR would develop the next logical step to the EZ-Rocket, a suborbital vehicle for passenger travel named Xerus.

The Xerus would be developed as a two-seater, Horizontal Take-Off, Horizontal Landing (HTHL) suborbital vehicle capable of reaching an altitude of sixty-two miles (one hundred kilometers), the acknowledged boundary of space. Passengers would experience several minutes of microgravity and 4 g on the way down. Space Adventures would sell the ride for $98,000 a ticket. Depending on funding, XCOR anticipated development would take eighteen months to the first test flights and a total of three years before the commencement of commercial passenger operations.

XCOR imagined other potential markets for Xerus, including the launching of suborbital science experiments such as materials processing, traditionally launched by sounding rockets. Microsatellites could also be launched into low Earth orbit via an expendable second stage attached to the Xerus vehicle. These would dramatically cut the cost and lead time for such launches.

With plans for developing a suborbital vehicle under way in 2003, XCOR was one of first companies, along with a number of rocket entrepreneurs, to proactively engage with and lobby Congress for clarification over which part of the FAA regulated the licensing of suborbital reusable vehicles. "The Commercial Space Launch Act of 1984 is clear," said Jeff Greason. "The law directs the secretary of transportation to license suborbital rockets and says that no permission from any other executive agency is required."

But the definition of "suborbital RLVs" was not clear, and licensing hinged on whether it should be regulated as a launch vehicle or an airplane. "When work started on regulating the industry, there was no such thing as a reusable launch vehicle," Greason said. "Nobody defined suborbital rockets anywhere." Having RLVs fall under the jurisdiction of aviation would make the licensing process so cumbersome that the industry would take decades to get off the ground. Eventually, regulation would come under the Office of Commercial Space Transportation (AST) at the FAA. XCOR submitted its FAA licensing application in November 2003 and received its license the year after, only the second company to be so licensed.

While XCOR was now poised to fully pursue its suborbital plans, a very different opportunity presented itself in 2005. Peter Diamandis, who had created the X PRIZE competition that was motivating much of the private spacecraft development, had moved on to his next ambitious venture. Along with entrepreneur and Indianapolis race car champion team winner Granger Whitelaw, Diamandis had just founded the Rocket Racing League. The league wanted XCOR to develop and build the first generation of Rocket Racers. These craft would race in the air equivalent to the Formula One races or, to the younger-generation sci-fi fans, a *Star Wars* pod race.

The races would be held across the United States, starting at the New Mexico spaceport in Las Cruces. They would feature several teams going head-to-head competition in a four-lap, multiple-elimination-heat format, closed-circuit "raceway in the sky." Pilots would view the race course via in-panel and 3D helmet displays, while fans would witness the real-time action on multiple large projection screens on the ground.

XCOR created the Rocket Racer based on the EZ-Rocket design and incorporated a next-generation airframe built by Velocity, a company recently acquired by the Rocket Racing League. Carrying a new XCOR 1,500-pound-thrust LOX/kerosene rocket engine, the single-pilot, single-engine Rocket Racer can reach a maximum speed of 230 mph. It has the ability to rapidly refuel, requiring on the order of five to ten minutes per pit stop. Its engine incorporates electric ignition that can be stopped and started in midflight as often as needed and is capable of three and a half minutes of intermittent engine boost.

In 2008 the Rocket Racer debuted before the public at the Oshkosh air show. It has since completed more than forty successful rocket-powered flights to date. For XCOR, the Rocket Racer was the next logical step to continue development on its line of increasingly more powerful rocket engines that would eventually be capable of launching into space.

In early 2008 XCOR announced a brand-new suborbital vehicle called the Lynx, aimed squarely at the space tourism market. A two-seater, HTVL suborbital vehicle, "the Lynx will offer affordable access to space for individuals, researchers and educators," said Jeff Greason. The Lynx will start off with flights reaching two hundred thousand feet, or sixty-one kilometers—short of the accepted boundary of space. The flight experience would rocket would-be tourists straight up at Mach 2 speeds to experience 4.25+ minutes of weightlessness at apogee, as they get treated to a breathtaking view of the Earth.

It will glide back down after experiencing 4-g forces during reentry. "Flying with us onboard the Lynx is more than just an hour-long whatever flight to the edge of space and back," explains Colonel Rick Searfoss, XCOR's chief test pilot. "The fact that you're up in the pilot seat as opposed to being in the back like a passenger in an airline, this is more like the 'Right Stuff' kind of experience." Searfoss knows a thing or two about the right stuff, being a retired astronaut and one of only seven people who have flown more than fifty rocket-powered flights.

"We have designed this vehicle to operate much like a commercial aircraft," Greason explained. Besides tourism, "future versions of Lynx will offer ever-improving capabilities for scientific and engineering research and commercial applications." XCOR plans to get the Lynx up and running by 2010 and offer flights several times each day. According to Greason, XCOR's Lynx will further "strengthen the financial and technical foundation for increasingly capable future spaceships for suborbital and orbital markets." In December 2008 XCOR successfully tested their new Lynx engine.

As part of XCOR's push to attract investors and actual passenger flights on the Lynx, they have partnered with Rocketship Tours, a luxury travel company headed by founder Jules Klar. Klar created $5-A-Day tours with Arthur Frommer in the 1960s, allowing affordable European travel.

As of December 2009, Rocketship Tours had booked more than thirty suborbital clients at $95,000 a ticket. It takes a $20,000 deposit to secure a spot on the list. The package includes five days in a luxury resort, medical screening, training on the spacecraft's subsystems and life support systems, and the suborbital flight itself. Each person who gets a ride on the Lynx will be seated side-by-side with the pilot, wearing a pressure suit, just like a regular test pilot. Suits are currently being developed by Orbital Outfitters.

XCOR's first suborbital reservation came from Danish investment banker Per Wimmer. "I am going to fly aboard the Lynx because I want to experience space from the front-row seat," Wimmer said in an XCOR press conference.

Aside from enlisting specialized travel agents to market their flights, XCOR is also working on a number of partnerships with spaceflight programs, including the revived Teacher in Space program. Early in 2009 XCOR awarded a free suborbital flight to Tuskegee airman Le Roy Gillead as part of their Legacy program, to thank persons and groups that have contributed to excellence in aviation.

In December 2009 XCOR announced that it signed a binding memorandum of understanding with the nonprofit Yecheon Astro Space Center in the Republic of Korea. This could potentially funnel $28 million in investment to XCOR to develop the next generation of the Lynx, capable of reaching an altitude over one hundred kilometers. With funding in place, the first prototype could roll off the production lines by 2011.

XCOR Aerospace is not the only company benefiting from a long line of SSTO RLV developments pioneered by people who had worked for DC-X and Roton. A number of the DC-X personnel have also been hired by another, more secretive vehicle developer, called Blue Origin, based in Seattle, Washington.

Blue Origin was founded by space buff and multibillionaire entrepreneur Jeff Bezos, the founder and CEO of Amazon.com. Incorporated in 2002, Blue Origin did not go public until 2003, and the company remains tight-lipped about its current development progress and schedule.

Jeff Bezos took an early infatuation with space and a tech background on a circuitous career path to end up in the private space business. He was born in Albuquerque, New Mexico, in 1964. A tinkerer from an early age, he dismantled his crib with a screwdriver while still a toddler. In high school he became hooked on computers and turned his parents' garage into a science lab. He won a trip to NASA Marshall Spaceflight Center in Alabama for writing a paper on the effects of microgravity on houseflies. At his graduation speech as valedictorian, he talked about space colonization. He was already dreaming of Gerard O'Neill's grand space settlements, of hotels, amusement parks, and yachts in space.

Bezos entered Princeton, intending to major in physics, but finished with a computer science and electrical engineering degree. His early career straddled the fields of computers and finance. Then he got a crazy idea about selling books over the Internet. On 4 July 1995 he took a gamble with his wife, Mackenzie, and left the security of the finance world to move to Seattle. They drove cross-country from Texas to the northern Pacific Coast in a 1988 Chevy Blazer while Bezos wrote the business plan for his new company on a laptop. A few days later they opened Amazon.com and set up shop in a two-bedroom house. Two years after Amazon.com went public, the Internet company's market value was bigger than its two biggest retail book competitors combined. Bezos would use his Internet wealth to reconnect with his dreams about space.

"Why did the founder of Amazon.com and a famous cyberpunk novelist [Neal Stephenson] ask for a tour of NASA Jet Propulsion Laboratory last February?" That was the opening line of a July 2003 *Wired* magazine article that tried to get to the bottom of Bezos's mystery visit.

Not knowing why an Internet multibillionaire would want to visit a world-renowned robotics laboratory, JPL gave the pair a VIP welcome, in the hopes of soliciting sponsorship. But contrary to JPL's robotics specialty, both men were talking about human spaceflight. "If we think outside the box, there's going to be a revolution," Bezos told the JPL staff. Nothing came of the visit except that he hired JPL engineers to populate his new Blue Origin manufacturing lab.

Shortly after visiting JPL, Bezos started buying up large parcels of property in the Texas desert east of El Paso. He now owns 165,000 acres of land in Culbertson County, at ComRanch north of Van Horn, Texas. Although his main headquarters and manufacturing facility still remain in Kent, Washington, outside Seattle, he is building a private spaceport to conduct his future space tourism business in Texas.

"We are building a vertical takeoff, vertical landing spacecraft that will take three or more astronauts to the edge of space," Bezos disclosed in November 2007 on the PBS TV Charlie Rose Show. He also shared his own intention to go to space. "People tell me who have been in space that it's a transformative experience," he enthused. Bezos had been thrilled to get a phone call from private space flight participant Charles Simonyi during the latter's visit to the International Space Station in early 2007. "I will go. I definitely will go," says Bezos. "I can't wait actually." However, he wants to go up on his own terms, on a Blue Origin vehicle called the New Shepherd.

Bezos's suborbital vehicle New Shepherd is another VTVL craft. It consists of a pressurized crew capsule atop a reliable propulsion system. The suborbital vehicle will be operated totally autonomously by onboard computers and be capable of lifting three or more passengers to 325,000 feet. Its current landing configuration has changed from a powered-down to a parachute-assisted landing. New Shepherd will use high-test peroxide and rocket-grade kerosene for fuel in its cluster of nine rocket engines, currently being developed.

The flight will take a maximum of ten minutes. Engine shutdown would occur at two and a half minutes into the flight, with those onboard experiencing micro-g for a total of three minutes. On the return segment of the flight, passengers would experience a maximum of 6 g. Bezos foresees flying at least once a week and is building the infrastructure on his private spaceport to conduct all components of the flight, including ground training and medical checks and procedures prior to flight. The projected timeline for human flight operations is currently planned for 2012, with the possibility of unmanned experimental flights in 2011. This timeline has slipped from original postings on their Web site.

After the FAA conducted an environmental assessment of Bezos's proposed Texas spaceport, Blue Origin obtained its experimental test flight license in September 2006. A few months later it flew its test vehicle, called Goddard, as a propulsion and control demonstrator. It reached a maximum height of 285 feet. The demonstrator looked vaguely similar to the conical shape of the DC-X and the Roton. As of December 2009, Blue Origin had selected three scientific projects from three universities to fly on their next test flight of the Blue Shepherd.

"We are not in any hurry because we are trying to build a very safe, well-engineered vehicle," Bezos explained. "Our motto is *Gradatim Ferociter*, 'Step by Step, Courageously.'" Although his main business target is space tourism, he confesses that he does not know how big the market is and is skeptical of market studies. "You don't really know until you do it," he further noted. "My passion is for space for sure, but I do think this can be made into a viable business. I think you have to be very long-term oriented. People who invested in Amazon for seven years would be horrified with Blue Origin."

Many other RLV companies have come and gone, while others are still trying to make that leap from page to hardware, but we have only scratched the surface. The best compilation of RLV work being done can be found in the quarterly report from the AST office of the FAA.

For most of the companies, money is the issue. And even fully financed RLV developers like Bezos still run into other technical and management issues along the way.

It remains to be seen which private companies will succeed in offering fully commercial suborbital flights and beyond. But as Bezos puts it, "I knew that if I failed I wouldn't regret that, but I knew the one thing I might regret is not trying."

7. The Ansari x PRIZE Launches an Industry

The best way to predict the future is to create it yourself.

Peter Diamandis

If Peter Diamandis's Greek immigrant parents had believed in omens, they might have made much of the fact that their son was born in 1961, the year that Yuri Gagarin became the first person to travel in space. But space was very far from their roots on the small Greek island of Lesvos. They had struggled through poverty and the horrors of World War II to put together their own vision of the future. His father earned a medical degree, met his future wife on a blind date, and moved to America to practice medicine in the Bronx, New York. There the vision unfolded. They built a comfortable life and a successful medical practice, moved to Long Island, and had a son. They imagined that one day he too would pursue a career in medicine and his life would unfold in the same orderly fashion. They failed to account for the moon landings.

A generation of youngsters watched the drama of the Apollo program unfold on their TV screens. In a great many young minds, the glorious accomplishments of the Apollo program suddenly made every vision of space seem more achievable. It certainly planted that seed in Diamandis. He credits Apollo for bringing him to the realization that his "mission in life was going into space and taking other people with me."

While his parents urged him toward a career in medicine, Diamandis wanted to be an astronaut. "It became sort of my personal, hidden mission. I wanted to be sort of a space explorer, entrepreneur." That mission is still very personal but not very hidden. Diamandis did earn his medical degree, but the influence of space on his life has been more pervasive and consuming. He has become one of the most high-profile individuals driving the private development of space. Though perhaps best known as the creator of the X PRIZE, the competition to launch the first nongovernmental, reusable spacecraft into space, Diamandis has been instrumental in the formation of a long list of space organizations and projects that have promoted private space efforts.

Diamandis's penchant for mobilizing people to support space causes began in his freshman year at Hamilton College in 1979. Disgruntled with the cancellation of the NASA-Department of Energy Solar Power Satellite study and the NASA mission to Halley's Comet, he persuaded nearly half of Hamilton's two thousand students to sign a petition opposing the cancelation. Although his efforts netted nothing more than a polite response letter, they whetted his appetite for space activism. A year later he transferred to MIT in the hope of getting more involved in space.

When he discovered that MIT had no space organization, he set about remedying the situation. "That's when the idea of SEDS came together." Students for the Exploration and Development of Space, or SEDS, came into existence when he persuaded two fraternity brothers and their girlfriends to sign his application to create a new student organization. The inaugural meeting drew thirty people, including Eric Drexler from the Cambridge L5 Society. The exhilaration of connecting with like-minded people and organizing them to a cause energized Diamandis.

At the end of that meeting, "I remember coming out of the Student Center and looking up at the star filled sky and feeling like 'Wow. This is really going to work!' When things click and you have that moment of realization that this could really be real. I really felt that SEDS could be a large and viable organization. It was great."

Diamandis kept up the momentum by persuading friends to open SEDS chapters at Princeton and Yale. In letters to *Omni* and *Astronomy* magazines, he announced the creation of the coalition, lamented the sorry state of the U.S. space program, and charged students to help make a difference. Large space groups, such as the Maryland Alliance for Space Colonization and George Washington University's SPHERE, signed on as SEDS chapters.

Although other space organizations already existed, the letters made SEDS appear to be a large, well-organized operation. The growing list of new chapters added to this impression. Only later did these new chapters realize that Diamandis counted each one of the letters he received in response to his magazine piece as a "chapter." As one long time Diamandis friend puts it, Diamandis was good at invoking the "entrepreneurial bluff mode" to take him places.

It was through SEDS that Diamandis connected with those individuals who would play key roles in many of his future projects. Two of the responses to his letters came from Todd Hawley, a student at George Washington University, and Robert Richards, at Ryerson Polytechnical Institute in Ontario, Canada. Both enthusiastic space advocates, they would play a large role in managing SEDS and later cofounding such Diamandis projects as the Space Generation Foundation in 1985 and the International Space University (ISU) in 1987.

Through the SEDS international founding conference in Washington DC in July 1981, Diamandis connected with Gregg Maryniak, a lawyer working at the Space Studies Institute in Princeton, New Jersey. Maryniak would later introduce Diamandis to Gerard O'Neill, serve on the faculty of ISU, and be instrumental in providing the inspiration for Diamandis's X PRIZE idea.

Within two years, SEDS had grown to be the largest student space organization in the world, with one hundred chapters in a dozen countries.

Studying medicine at Harvard was part of Diamandis's plan to become an astronaut, as was the temporary interruption of those studies to earn undergraduate and graduate degrees in aeronautical and astronautical engineering at MIT. Busy with space activities during his college years, he had the opportunity to talk with several career astronauts, such as Byron Lichtenberg. From them he got a clearer picture of his chosen career path. "Your chances are one in a thousand for getting selected [in the astronaut corps]," Diamandis explained. "And even when you get selected, you might, if you're lucky, get to fly into space once during your entire career as an astronaut. It was not my vision of space."

If he really wanted to make his dream of space travel happen, Diamandis realized, he would have to create the spaceflight opportunity for himself. Which is pretty much what he set about to do.

In 1985, with Richards and Hawley, Diamandis founded the Space Generation Foundation, to give students and young space professionals a voice in space activities. A year later, working through the foundation, the same trio created the ISU, now a multimillion-dollar institution based in Strasbourg, France. ISU would have a far-reaching impact on the space industry. As with his other projects, Diamandis and his collaborators would attack this venture with unbounded energy and zeal.

After a founding conference for ISU at MIT in 1987, the trio set about recruiting an international faculty of experts and persuading space agencies from the USSR, China, Japan, India, and other countries to send their best and brightest to the first ten-week summer program the following year. Diamandis had no problems cold calling space experts and potential sponsors to get what he wanted. Most of the faculty, experts, and financial backers who joined the Diamandis bandwagon report the same initial encounter. Peter had the gift of persuading people to rally to a cause. An eloquent speaker, he was capable of connecting with individuals in one-on-one conversations or inspiring a fully packed hall of space followers.

As a result of that dynamism, the ISU was up and running within one year and quickly established itself as the premier institution for training space professionals. Its graduates, from over ninety-five countries, have since occupied increasingly higher positions in both government and private space industries, as astronauts, heads of space agencies, and science experts around the globe.

For most graduate students, still in their twenties, launching an international university would have been a crowning achievement. For Diamandis, success only meant that his schedule opened up, giving him time for other challenges, for the next big space thing—time, for instance, to cofound Space Adventures and Zero-G Corporation, which offered parabolic flight opportunities. Time to play a role in other start-up companies, such as Starport.com in 1999, a space news portal that was bought out by rival Space.com, and International Microspace in 1989, intended to provide low-cost launch services.

Diamandis's entrepreneurial juggernaut eventually gave rise to its own manifesto, "Peter's Laws," rules of the road that defined both an operating strategy and a personality. They are offered here with permission.

Peter's Laws™

1. If anything can go wrong, Fix It!! (To hell with Murphy!)
2. When given a choice — Take Both!!
3. Multiple projects lead to multiple successes.
4. Start at the top, then work your way up.
5. Do it by the book . . . but be the author!
6. When forced to compromise, ask for more.
7. If you can't beat them, join them, then beat them.
8. If it's worth doing, it's got to be done right now.
9. If you can't win, change the rules.
10. If you can't change the rules, then ignore them.
11. Perfection is not optional.
12. When faced without a challenge, make one.
13. "No" simply means begin again at one level higher.
14. Don't walk when you can run.
15. Bureaucracy is a challenge to be conquered with a righteous attitude, a tolerance for stupidity, and a bulldozer when necessary.
16. When in doubt: THINK!
17. Patience is a virtue, but persistence to the point of success is a blessing.
18. The squeaky wheel gets replaced.
19. The faster you move, the slower time passes, the longer you live.
20. The best way to predict the future is to create it yourself!

(Copyright 1986 by Peter H. Diamandis. All rights reserved. Laws 14 and 18 by Todd B. Hawley.)

The idea for the X PRIZE competition took root in Diamandis's head in 1994 after reading Charles Lindbergh's book, *The Spirit of St. Louis*, which recounts Lindbergh's methodical strategy for winning the Ortieg Prize, awarded for the first nonstop flight from New York to Paris.

The book was a present to Diamandis from Gregg Maryniak. Ever since their involvement with SEDS, the two had become good friends, fellow space advocates, and business partners. An avid pilot since he was sixteen, Maryniak took Diamandis up in a Cessna 172 on a crisp fall day near Princeton, New Jersey. A week later he found a copy of *The Spirit of St. Louis* in mint condition and bought it with Diamandis in mind. When he was a freshman in high school, the book had inspired Maryniak to take flying lessons. He thought it might do the same for Diamandis, who had begun pilot training years early. Diamandis would eventually get his license years later.

When Diamandis finally got around to reading the book, it had an impact far beyond anything Maryniak intended. Diamandis could not put the book down; he read it twice over, making notes in the margin as he went. Substitute "spaceflight" for "flight," and the seventy-year gap between Lindbergh and Diamandis melted away. Lindberg's thoughts might have been his own. There was a remarkable similarity in the way both men strategized to turn a dream into reality. No small wonder that Diamandis took the book to heart—it had shown him the way to accelerate private spaceflight development. He would create the first suborbital spaceflight prize.

The concept of using prizes to jump-start innovation was hardly new. From advancing aviation to the precise detection of time, prizes have played a major part in advancing radical technology throughout history. In 1714 Britain's Longitudinal Act created cash prizes for the first marine chronometer, later won by Englishman John Harrison in 1735. In the eighteenth century Nicholas Leblanc's process for producing soda from seawater was also spurred on by the engineering prize of 100,000 francs offered by the French Academy. It would be considered one of the key chemical engineering inventions of all time. During more recent times, the Nobel Prize–winning physicist Richard Feynman offered prizes for developments in nanotechnology. The annual Feynman Prize, given by the Foresight Nanotech Institute, still stimulates developments in that field.

According to Diamandis, "Prizes are most effective when progress is blocked and where market forces, government, and non-profits cannot readily solve a problem. They mobilize entrepreneurs to achieve breakthroughs."

Maryniak himself had written a paper for the 1993 Space Studies Institute Conference about using a Lindbergh-like event to help human spaceflight development. In Maryniak's article, "When Will We See a Golden Age of Space Flight?," he cited the hundreds of incentive prizes offered in the twentieth century and their role in creating what is now the multibillion-dollar commercial aviation industry. Several prizes were offered in Europe for the fastest, highest, and longest-duration flights within the continent.

The amount of prize money was significant, and the aerial requirements demanded were considered close to impossible at the time they were offered.

Europe's progress in aviation far outpaced that of the United States at the time. Albert F. Zahm, then head of the revived Smithsonian Aeronautical Laboratory, was dispatched to Europe to understand their rapid progress in response to these prizes. His report later became the basis for the creation of the National Advisory Committee for Aeronautics, or NACA, the precursor to NASA.

Diamandis did his own research on Lindberg's 1927 winning of the $25,000 Ortieg Prize (equivalent to roughly $300,000 in 2010 dollars). His prize-winning flight captured the imagination of the common man and created the business boom the aviation industry needed. Once the public began to believe that flying could be an experience for everyone, the result was increased demand, lower prices, and better performances.

It was the perfect formula for a spaceflight prize. Diamandis would name it the "X PRIZE"—the X serving as a placeholder for the name of the benefactor who would give the needed prize money. He also settled on $10 million as the prize amount, a nice round number that is also represented by "X" in Roman numerals. Now he was ready to recreate the same golden age of aviation in spaceflight through a private spaceflight competition.

Diamandis wasted no time telling Maryniak of his wild idea. "I was the first guy to tell Peter he was crazy," Maryniak recalled. But Maryniak soon recanted, knowing full well that this might be the solution to bringing down the wall that was stopping all progress." The wall, quite simply, was the "premise that only governments could do it [spaceflight]. We needed a prize like the Ortieg Prize to get human spaceflight unstuck."

Over the following year, Diamandis and Maryniak worked on the X PRIZE project in Diamandis's Rockville, Maryland, home during Maryniak's monthly visits to the Washington DC area, while working for the Futron Corporation. In the process, they recruited to the project former astronaut Byron Lichtenberg, who had retired from NASA and was a Southwest Airlines pilot.

Diamandis wanted an achievable stepping stone to private space exploration and thought that a suborbital flight, which uses only about 1.7 percent of the energy needed to launch an orbital flight, would be a reasonable goal for nongovernmental entities. The trio would eventually set the X PRIZE's requirements to include building a manned space vehicle, without the help of government funding, to be able to launch three people to an altitude of one hundred kilometers successfully and do it again within two weeks.

With the details of the prize set, Diamandis went about creating a non-profit foundation to raise the money. He chipped in a few thousand dollars himself, and a few thousand more came from Tom Rogers, who also provided ISU's first seed investment. Through mail solicitations, he raised about $10,000 to $20,000, enough to keep things rolling. However, by late 1995, Diamandis realized that raising the $10 million prize was "not going to happen doing it this way. [I] needed to find a community of people who could support and underwrite the X PRIZE."

Ironically, the community that he would find to support the X PRIZE was the same community that had supported Lindbergh—St. Louis, Missouri. As Maryniak would later characterize the attitude in St. Louis, "they talk about Lindbergh like it happened last Tuesday."

The impulse to support grand dreams still lingered in St. Louis. That explained the reception Diamandis got when Doug King from the St. Louis Science Center arranged for him to speak to Alfred "Al" Kerth, who headed an organization called Civic Progress in St. Louis. Kerth was at that time also senior vice president for Fleishman-Hillard, one of the top public relations firms in the country. The X PRIZE Foundation needed a permanent home base, Diamandis explained. What better location for the foundation than the city that had helped Lindbergh win the Ortieg Prize?

"In the middle of my presentation," Diamandis recalls, "Kerth stands up and says, 'I get it! It's great! We need to take this to St. Louis.'" As if to confirm just how much he really did get it, Kerth called Diamandis later that same night to say that he'd found a way to fund the prize. "When Lindbergh needed funding, he went to nine St. Louis residents and got them to contribute $25,000, and they called themselves 'The Spirit of St. Louis.' We're going to find 100 people in St. Louis to raise 2.5 million dollars."

The group would later be called "The New Spirit of St. Louis," and its contributions would jump-start operations, pay for a kick-off event, and help raise the prize money. And so King and Kerth became Diamandis's unsung heroes the way Harry Knight and Harold Bixby were Lindbergh's financiers and constant support in developing the *Spirit of St. Louis* for the transatlantic flight in 1927.

Diamandis continued with the Lindbergh script by holding a meeting for prospective investors at the Racket Club, the same venue Lindbergh used almost seventy years before. More than twenty-five supporters signed up, providing enough money to finance the official gala announcement of the X PRIZE on 18 May 1996 under the historic arch in St. Louis. The guest list for that event gave some indication of the support building behind the prize. NASA administrator Dan Goldin, famed aerospace designer Burt Rutan, and some twenty astronauts gave the X PRIZE a high-profile launch.

Unfortunately, all that publicity did not yield the additional supporters needed. X PRIZE operations continued as a bare-bones operation run by volunteers. From 1996 to 1997 Maryniak worked for the foundation part-time and for free, commuting to DC or St. Louis from his home in Pennsylvania. Diamandis operated from the basement of his home in Rockville, Maryland. In 1997 he was assisted by a summer intern, Eric Anderson, whom he had met at a NASA Academy held at Goddard Spaceflight Center the previous summer. Anderson designed X PRIZE's first Web site. The number of teams registered to compete for the prize grew to fourteen. In the fall of that year, Diamandis moved to St. Louis.

If the financial well-being of X PRIZE operations over the next few years were plotted on a graph, it would reveal the manic hills and valleys of excitement and high anxiety. While creative financial arrangements and additional partners provided surprise funding at critical moments, the specter of bankruptcy continually dogged the foundation.

Part of the financial puzzle fell into place when Diamandis got advice from his friends Bruce Krezelski and Bob Weiss. Bob Weiss, a Hollywood producer of such movies as *Blues Brothers* and the *Scary Movie* series, would later join Diamandis and Maryniak on the X PRIZE team. For now, what he wanted to suggest was that making a suborbital rocket launch was similar to a hole-in-one golf competition. Organizers paid a premium to cover themselves in the event someone sank a hole-in-one. The X PRIZE Foundation needed to find an insurance company willing to cover the $10 million prize in the event someone won the competition.

Diamandis took their advice. He identified three companies willing to consider coverage. For their due diligence, the insurers talked to the major aerospace companies, Boeing and Lockheed, these being the private entities that would be most able to win such a challenge. Were they going to compete, the insurers asked? The answer was no. Did they believe anyone could meet the challenge in the allotted time frame? They got another no. So the insurers came on board. The end date for the prize was set for 17 December 2003, the hundredth anniversary of flight. (The deadline would later be extended to 1 January 2005.)

A series of watershed financial moments moved the X PRIZE closer to reality. One of the biggest such event came innocently enough when Diamandis arrived at his St. Louis office one day to find a note that read, "This guy called. Called regarding a donation." Diamandis remembers writing on the pad, "We don't make donations," and giving it back to the secretary. She called again to clarify that they wanted to *make* a donation to the foundation.

The call turned out to be from First USA, a bank headquartered in Delaware. The bank representative explained that he had read about the X PRIZE St. Louis Gala and about how writer Tom Clancy had made a $100,000 pledge. Clancy had been so taken with the gold medallions that the X PRIZE Foundation gave to the Spirit of St. Louis members that he asked how he could get one. When told that they were given to each $25,000 member sponsor, he said, "I'll take four."

Impressed, the banker thought that an X PRIZE credit card would "really be cool." First USA signed a deal in early 1998 that would eventually pay the foundation $5 million over the next six years—half the money needed for the prize.

"On the strength of that [deal], I took the wild chance of moving my family" to St. Louis, Maryniak recalled. Maryniak began working full-time on the X PRIZE as executive director of the foundation. It would be Maryniak's job to find and grow the New Spirit of St. Louis membership and the Senior Associate program for the public, and to manage recruitment of the X PRIZE teams. He set up shop in an office inside the St. Louis Science Center. For one brief moment, the financial picture looked rosy.

Diamandis and Maryniak had thought it would be smooth sailing after the initial influx of investors, but the financial struggle continued. Between 1999 and early 2001, the foundation was flat broke, propped up only by some yearly financial miracle. Time and time again the New Spirit of St. Louis organization would help them out through an influx of new members and their accompanying dues.

"We had many, many, many near death experiences as an organization." Maryniak is quick to cite the parallel comparison to Lindbergh's St. Louis backers, who amazingly "backed this twenty-five year old, air mail pilot who had this crazy idea" of using a single-engine plane and a single crew member for the transatlantic flight. "We have the same blind support from the 50ish St. Louisians that backed the X PRIZE Foundation. The *Spirit* was still there. They had a special civic pride and a belief in taking risk for technical innovation that I have not found in any other community," Maryniak said.

Their support and more would be required. Diamandis negotiated a payment scheme with an insurance company for the other $5 million in prize money. It required a $50,000 monthly payment with a large balloon payment at the end. "I remember there would be these $50,000 Fridays coming up where it would be Monday and I have to raise 50K by the end of the week. If I didn't raise that money, everything that we had worked on would be gone. The prize would be over. You can imagine the incredible stress in doing that," Diamandis recalled.

The St. Louis Science Center, the Danford family, and Blast-Off, an idea lab company, were also major players in keeping X PRIZE afloat during those years. Maryniak was already thinking of finding a part-time job. Diamandis had relocated to Pasadena to join Blast-Off as part of a sponsorship deal for the X PRIZE.

In March 2001, on the verge of going out of business, they had a "powwow" in Pasadena to figure out what to do. Maryniak was so strapped for cash that on the way back from the meeting, he was worrying about how to pay for his daughter's $3,000 orthodontic bill. Back home he found his car badly damaged by a hailstorm. He drove to the nearest State Farm insurance center for the payoff and got enough cash to pay the bill, with $17 to spare. "It was dreadfully tough. We were dead broke so many times." Maryniak recalled.

But once again the Lindbergh connection came through, this time in the guise of the famous aviator's grandson. Eric Lindbergh had signed on early as one of the X PRIZE's critical supporters. In 2001 Lindbergh created a sponsorship deal with the St. Louis Science Center to reenact his grandfather's historic transatlantic flight using a modern airplane. The successful flight, occurring the following year, raised about $1 million for the foundation. "It was the only time during the x PRIZE history where I knew where the money was coming from over a six-month increment," Maryniak noted.

The Lindbergh money was another of those last-minute financial reprieves. It came at a critical moment, in a post-9/11 world, when potential airline sponsors had disappeared. But like all of the foundation's financial reprieves, it was less the result of good fortune than of dogged persistence. Number 17 of Peter's rules counts persistence as a blessing, a vehicle for creating good fortune. Diamandis's persistence was about to pay off in a big way.

A 2001 article in *Forbes* magazine on the world's wealthiest women under the age of forty featured Iranian American businesswoman Anousheh Ansari. The article pointed out that her dream was to fly a suborbital flight into space. "When I read that article, I remember exactly where I was," Diamandis recalled. "I was in my apartment in Santa Monica. I said, 'This is her.' I knew it instantly. This was my benefactor."

As soon as he finished the article, he picked up the phone. After a few calls, he connected with Anousheh's former private secretary. "Well, Anousheh's on vacation in Hawaii . . . have to wait . . . in a month," said the lady on the line. "No, no, I've got to meet her now." Diamandis explained who he was and what he was doing. She listened patiently and in the end said, "I know Anousheh would love to talk with you."

Diamandis ended up sending Ansari a FedEx package, then met with her as soon as she returned from vacation. He explained to her the concept behind the x PRIZE: how such competitions can lead to radical breakthroughs in technology, and how a small group of people with passion can harness the power of public interest, the entrepreneurial spirit, and the drive of innovators—very much as Anousheh herself had done in her technology business—to solve some of the greatest challenges facing the world today. The x PRIZE wanted to do just that with private space travel.

During his presentation, Anousheh looked at her husband, Amir, smiled, and said they would do it. She would later recall how impressed she was with Peter's determination. She had never heard of Diamandis prior to the meeting but decided to take the risk because of his passion and his belief in what he was doing. "It was not just a hobby but a lifetime calling." The bottom line: she saw herself in Diamandis. When asked about the huge risk she took on the X PRIZE, she said, "We are a family of risk-takers. We are entrepreneurs."

The X PRIZE became the Ansari X PRIZE in 2004. The multimillion-dollar infusion allowed Diamandis to pay off the insurance policy balloon payment and continue operations. Other large sponsors followed: Kevin Kelkoven, former CEO and chair of JDC Uniface, became a $1 million supporter. 7Up paid $700,000 for the advertising privileges. Champ Car Racing became the presenting sponsor prior to the Mojave launches. By the end of the competition, Diamandis would end up spending every dollar he had received.

Shortly after receiving the Ansaris' initial funding, Diamandis hired former NASA Glenn Research Center (formerly called NASA Lewis Research Center) employees Ken and Gretchen Davidian. The Davidians were a dynamic duo who had previously worked for the ISU. Ken, a fast-talking, lively aerospace engineer with an infectious laugh, complemented his wife's more reserved but pleasant disposition. They were the type of industrious and dedicated people the foundation needed to get things done despite a lack of basic material support and less than normal compensation. And there was a lot to get done. Diamandis was unable to prioritize their initial one-hundred-item to-do list since everything on it was equally important. Gretchen became director of communications and Ken director of operations.

Ken's major task was to set up the formal registration process for the X PRIZE teams, which did not really exist prior to his arrival. This was surprising, considering that there were already twenty-three registered teams. In an effort to be more inclusive in the beginning, Ken explained, "they kind of just opened up the door" to everyone who could pay the $1,000 registration fee.

Of course, this opened up the competition to several unusual entrants, including somebody who wanted to use antigravity for his propulsion system. "I had to write a letter to him saying, 'I'm sorry, we can't let you in,'" Ken recalled. Eventually he would finalize the registration forms and criteria, work out the team agreements, and start gathering data on their development status.

Of the two dozen or so registered teams, only about a third had crossed over from design stage to actual hardware. Everyone had the same problem of not having enough seed money to enter hardware development. As Ken puts it, most of them were funded by the "Three Fs: friends, family, and fools." Most didn't have the business knowledge and financial savvy to take it to the next funding-source level.

Although budget constraints confined their communications with most teams to phone and e-mail correspondence, Ken and Maryniak began visiting teams in July 2002. "I was very enthusiastic," Ken said. It turned into a two year adventure, as they visited teams in Washington DC, Texas, California, Ontario, and England. They never knew quite what to expect on these visits. Hard-headed investors, eccentric engineers, obsessed space addicts — the X PRIZE had galvanized a wide swath of individuals to lend their talents to the competition.

Advent Launch Services

One trip took Ken Davidian and Gregg Maryniak to Houston. There Jim Akkerman spearheaded one of the early entrants to the competition, Advent Launch Services. Akkerman was a former NASA employee whose decades of service started in pre-Apollo years at the Johnson Space Center. "He received royalty checks on a monthly basis for the design of the blood pump for the DeBakey artificial heart," Ken recalled. "So he was getting his patent checks and pumping it into building this rocket. He was using his first stage as his test stand. He set up his full-scale mockup of his first stage propulsion system up on these cinder blocks. Gregg and I went and drove up to his house, and it's right there in the garage."

Akkerman would conduct tests in a rice field outside of Houston. He sent the X PRIZE a video of his first engine test. "You would see fire at the base of the thing, and he would go running and tweak something and run away." Ken recalled. The full-scale vehicle was designed to be a single-stage rocket that would take off vertically from water and glide back horizontally to land in the water like a seaplane.

Akkerman had originally teamed up with another outfit called the Civilian Astronaut Corps (CAC) to develop the Mayflower Expedition, a spacecraft to take paying passengers on a suborbital flight, which was slated to fly by 2000. CAC and its plans for the Mayflower eventually folded. In June 2004, after two successful engine tests, fire damage due to an overheated oxygen line essentially took Akkerman out of the race.

Kelly Space and Technologies

Another visit took Ken and Maryniak to Kelly Space and Technologies (KST), another of the early entrants. Founded by Michael Kelly and Michael Gallo, KST was building the Eclipse Astroliner, a tow-launch, horizontal landing vehicle designed for the low Earth orbit satellite market. Kelly had twenty-five years of experience working in the aerospace business for Northrop Grumman and TRW. The team had been awarded a Motorola contract valued at $89 million and had already demonstrated its tow-launch system back in 1998, using a modified F-106 in cooperation with NASA Dryden. Considered to be a promising contender in the early years, the business end never panned out and the program went belly-up long before the prize was awarded.

Starchaser Industries

Steve Bennett from Cheshire, England, led Starchaser Industries, a contender from the United Kingdom. Largely self-taught, Bennett started out in amateur rocketry before working as a lab technician and serving in the British army. He set up Starchaser in 1992 as an experimental rocket program for cheap access of scientific payloads to high altitudes. Bennett and his team successfully launched a six-meter rocket called Starchaser 2 in 1996, making it the largest private civilian rocket to be built and flown in Europe.

As an early entrant for the X PRIZE, Bennett developed a two-stage, vertical take-off, parachute landing spacecraft called the Thunderbird. By 2001 Starchaser had launched a reusable rocket called Nova capable of carrying a three-person payload and boasting the largest rocket ever to be launched from British soil.

In 2002 some unusual circumstances prompted Ken to visit Bennett's manufacturing and assembly plant. Bennett had sued the BBC for publishing an online article describing his space capsule as a "converted cement mixer with sheets of hardboard and a few computer joysticks." Bennett claimed that such a characterization cost him potential sponsorships. But the BBC refused to take the article off its Web site.

Starchaser's attorney contacted X PRIZE, asking for an expert to come over and attest to the fact that their capsule was not a cement mixer but a legitimate capsule design. Having worked for NASA, Ken fit the bill. He flew to Manchester, checked out the capsule, and declared that "Yup, looks like a capsule, not a cement mixer." He wrote a report to that effect, and Bennett eventually won the suit and forced the BBC to remove the article.

The Starchaser team was confident it had a chance at the prize until it essentially ran out of time and the funding to take it through the home stretch in 2004.

Pioneer Rocketplane

When X PRIZE made its first big gala event announcement at St. Louis in 1996, a few competing teams had already unofficially registered. Present at the event were Pioneer Rocketplane representatives, eager for the challenge. Pioneer Rocketplane had been created by Robert (Bob) Zubrin, best known as the founder of the Mars Society and vocal supporter for Mars exploration; Mitch Burnside Clapp from the Air Force Research Lab; and Charles "Chuck" Lauer.

Lauer was essentially the company's front man and Vice President of Marketing. Tall with longish silver hair and beard, he seemed a better fit at a Renaissance festival rather than hob knobbing with aerospace engineers at space conferences. Lauer came from the real estate business but with a keen interest in space. He was a believer in space as the future of commercial development and thought Gerard O'Neill's grand visions for human exploration as "effectively a real estate process." The private sector, not just big government, should be developing this real estate and creating the associated economic opportunities.

In February 1995 Zubrin gave Lauer the opportunity to act on those beliefs when he asked him to participate in "this little suborbital space plane project." Zubrin and Clapp were already working on the Black Horse and Black Colt projects for the Air Force, and commercial prospects for private launch companies had never looked better.

In 1995 the telecommunications boom was in full swing. Mega-telecommunications companies like Iridium and Teledesic were fueling the market for big launch systems, with plans to launch more than a thousand communication satellites. So that same year, Lauer, Clapp, and Zubrin put together the business plan for what would be called Pioneer Rocketplane.

The launch system would consist of a reusable first-stage plane and an expendable second stage to take satellite payloads to orbit. They attracted some angel investors on the prospects of becoming a satellite launch service business. In 1996 they signed on their first six-figure investor, George

French from Wisconsin. French had also invested in Kistler Aerospace prior to coming on board. With the new seed money in place, Zubrin quit Martin Marietta to be full-time CEO, and Pioneer set up shop in Colorado. Lauer kept his day job in Michigan but began putting in more and more time on Pioneer.

Lauer knew Diamandis from the conference circuit, so when X PRIZE debuted in 1996, "it was an easy consensus. It does not hurt us to register for this," recalled Lauer. "We were building the vehicle anyway." Even though the spaceplane was planned as a much bigger and higher capability vehicle, the reusable first stage would fly a suborbital profile.

A few NASA contracts in 1997 helped Pioneer advance progress on the vehicle design, as did additional private investors. However, by 2000, the satellite communications industry had gone bust, leaving Pioneer with no "addressable market except for suborbital tourism." They began to downscale their launch vehicle significantly into a Lear jet configuration with four seats capacity.

In 2001 Pioneer was already in "survival mode" trying to figure out what they were going to do. The team almost folded, but instead changed their business plan to concentrate on winning the X PRIZE. Survival depended on obtaining tax credits from the state of Oklahoma, based on setting up operations at a spaceport the state was promoting at the old Clinton-Sherman Air Force Base. However, they did not anticipate that the tax credit certification would take two years, or that fourteen other companies would compete for the financing.

When the tax credits were approved on 31 December 2003, the name of the company was changed from Pioneer Rocketplane to Rocketplane Ltd. and its operations were set up in Oklahoma. At peak staffing, sixty-plus people worked on the spaceplane, but there was simply not enough time to get it operational during the X PRIZE timeline.

TGV Rockets

One of the teams working to get a piece of the Oklahoma tax credit was TGV Rockets. Ken met founder Pat Bahn during the July 2002 trip to Washington DC, and characterized him as a "strong, intense, no nonsense guy." Bahn certainly "had thought of what the markets might be, outside the plain old take-three-people-up-for-a-joy-ride kind of market," Ken noted.

Bahn came from the computer and Internet industries and was more business savvy than most competitors. He was keen on exploring suborbital flight applications other than space tourism, such as remote sensing, scientific payloads, and military reconnaissance. He made it his first priority to meet and line up potential clients from the science, government, and military sectors as proof of market demand to attract financial investors.

Bahn founded TGV in 1997. TGV stood for "Two Guys and a Van." The name pretty much summed up the company's goal and philosophy: Build a suborbital system portable and uncomplicated enough to be operated and launched by two guys, using off-the-shelf technology and a command center as big as a van, and you have a chance at bringing down the cost of launching payloads to space. TGV's launch vehicle, named Michelle B, used vertical take-off and vertical landing configuration with an innovative "shuttlecock" dive brake design for increasing drag at high altitude and reducing speed for a rocket-powered landing. In an age when a veritable tsunami of annoying acronyms flood the space scene, one would reasonably deduce that the name Michelle B might be different—perhaps a tribute to a close family member—but no. It stands for Modular Incremental Compact High Energy Low-cost Launch Example. Maxwell Smart would have been proud.

Although TGV's timeline was still in its start-up stage by the time the X PRIZE was awarded in 2004, Bahn would eventually win contracts from the U.S. Naval Research Lab, DARPA (Defense Advanced Research Projects Agency), and the U.S. Air Force. He would move his team, mostly former DC-X engineers, to a four-thousand-square-foot facility in Oklahoma, where he is now concentrating on a manned remote sensing system for scientific and military purposes.

Armadillo Aerospace

John Carmack registered Armadillo Aerospace late in the game. Well known in the computer gaming industry as the multimillionaire software developer of the video games Quake and Doom and cofounder of id software, Carmack did not fit the profile of the typical space entrepreneur. "I did not fall into that category," Carmack explained in a 2008 interview. He stumbled on his current passion due to a "random set of circumstances" that led him to back and financially support two rocketry teams vying for another smaller space race called the CATS Prize.

The CATS Prize, short for "Cheap Access to Space," was a $250,000 challenge sponsored by the Space Frontier Foundation and the Foundation for the International Non-governmental Development of Space (FINDS) and financed by telecommunications mogul Walt Anderson for the first team to launch two kilograms of payload to two hundred kilometers or higher by 8 November 2000.

When Carmack looked at CATS, only a year remained until the deadline, but he got drawn into the community of competitors. He interviewed all of the teams and eventually funded two of them, JP Aerospace and the Sorak team. Out of fifteen competitors, these teams were the only ones that had test programs. "Even if you take the most pragmatic plan possible, there are still a lot of things that can go wrong," Carmack stated. "It was my foot in the door. It was my learning phase. I grabbed all the books and I read everything, learning the ropes of what would be necessary to do something here." In the end, nobody won the prize in the time allotted, but the resulting learning process for Carmack led to the creation of Armadillo Aerospace on 1 January 2000.

Carmack started recruiting people in the Dallas area with experimental rocketry background. He went to the local rocketry club and asked the president if any of their members were involved in building their own engines. He ended up with essentially volunteers, half of the team coming from the rocketry club and half from "as-needed situations"—mainly family and acquaintances needed for the odd job here and there. (Years later, he still has his core team.)

Having watched one of the CATS teams develop their rocket Carmack then set about going through the same process. "They build a rocket for a year, go out into the desert, they press the button and hope it doesn't blow up. It really rarely works right."

He wanted to follow the same "rapid iterative process" he used in developing software and apply it to his rocketry business. "The background that I came from in software is you compile and test maybe a dozen times a day. It's a cyclic thing where you try to make it right but much of the benefit you get is in the exploration of the process, not so much plan it out perfect, implement it perfect for it to work. It's an iterative process of exploring your options." Carmack taught himself aerospace engineering and became one of Armadillo's principal engineers for the project.

Armadillo officially registered for the x prize in October 2002 when Carmack was sure the prize was funded. He then short-circuited his own strategy for building more intermediate vehicles and immediately went directly to a three-man vehicle testing phase. Ken Davidian would learn about Carmack's progress through his frequent blog updates on the Web.

Armadillo's vehicle design was unconventional, to say the least. It used a single-stage rocket in vertical liftoff configuration using attitude controls and returned using a ballistic trajectory with parachute deployment during reentry. The capsule finally landed on a crushable nose cone.

By 2003 Armadillo had already tested a manned single-person hovering vehicle and conducted numerous engine tests of bipropellant rocket engines, all on a shoestring budget of just over $1 million. By June 2004 Carmack and his team had successfully demonstrated a computer-controlled vertical take-off and landing flight to 131 feet, using its prototype x prize vehicle, making them only the third unmanned rocket in the world to do so. But a crash of the prototype vehicle in August that year wiped out Carmack's hopes of fielding a fully operational rocket to compete for the prize by the end of the year.

Canadian Arrow

Two of the more promising teams building hardware for the x prize were Canadian Arrow and the Da Vinci Project, both from Ontario, Canada. Ken Davidian met both Brian Feeney of the Da Vinci Project and Geoff Sheerin of Canadian Arrow on a visit to Canada with Maryniak in 2002.

The Canadian Arrow team from London, Ontario, led by entrepreneur Geoff Sheerin, was building a fifty-four-foot-long, two-stage vertical take-off rocket with a parachute water landing. Resembling the classic v-2 rocket, the second stage doubled as an escape system for a three-person crew. In April 2002 they trucked a full-scale mockup of the spacecraft from Ontario to Manhattan's Rockefeller Center as part of x prize's marketing push and Eric Lindbergh's announcement to recreate his grandfather's historic New York-to-Paris flight.

By June 2003 they had already chosen their first six astronauts after a global recruitment process. All six candidates had either aerospace or military backgrounds and would train for their planned two-person test flights. Later, the team would successfully conduct a drop test of their unmanned crew capsule in August 2004, but like many of the other competitors, they were outpaced by other deep-pocketed teams.

Da Vinci Project

The Da Vinci Project proved to be the underdog that tried to race the winner to the X PRIZE finish line in 2004. Its founder, Brian Feeney, could not be described as anything less than a driven man. In 1996 he was working in Hong Kong as a designer for Brita, the water filtration company. He learned about the X PRIZE from a newsstand, and from that moment the idea of building a rocket consumed him.

He started contacting aerospace companies and ordering supplies, eventually returning to his hometown in Toronto, Canada, after six years abroad. He recruited young engineers from the University of Toronto's Aerospace Institute, who eagerly volunteered for a chance to build a Canadian-made spacecraft. By the end of the competition, the project was billed as the largest volunteer-run technology project in Canadian history. Feeney managed to solicit in-kind donations and materials for his rocket and convinced the Da Vinci Polytechnic Institute to donate office space. He officially registered the project for the X PRIZE in the summer of 2000.

It was not unusual for Feeney to take up an idea and run with it. His enthusiasm for space started while watching one of the NASA Gemini missions on TV in the 1960s. Soon he was building and launching homemade rockets from his backyard. His official Da Vinci Project biography reports a harrowing rocket-obsessed youth. "We had our own Canadian 'October Sky' Program with one memorable launch resulting in a mini mushroom cloud in the local school yard." He almost burned down his family garage while building a flamethrower and had already measured out the hole he planned to cut in the attic of his boyhood home for an observatory before his parents put a stop to the project. He would construct his own photography lab for developing his astronomy photographs and won awards for building homemade eight-inch telescopes.

By age twenty-three, he was already the CEO of a publicly traded company. In the 1980s he had a successful life support systems company, for which he had several patents to his name. It was technology he would implement in the spacecraft's life support system. But the business eventually dried up, and he shifted gears, landing half a world away at Brita.

Feeney's launch system design was one of the most unconventional in the X PRIZE competition. The bullet-shaped spacecraft Wildfire, with its sixteen-window cockpit, was inspired by the B-29 bomber. Wildfire would be lifted by the biggest helium balloon ever manufactured to a height of eighty thousand feet, where it would ignite using a solid-liquid hybrid rocket. The motor and capsule would return using two separate parachutes. This parachute recovery had been simplified from an earlier version that deployed an innovative conical balloon acting as a heat shield.

Feeney would make incremental steps between 2001 and 2004, including engine and guidance testing and unmanned test flights of the spacecraft in Mojave and White Sands. On 27 July 2004 Burt Rutan announced that Scaled Composites had scheduled its x prize launch for 29 September. That same day Feeney announced the unveiling of his vehicle. On 5 August 2004 the Da Vinci Project officially rolled out to a great press show. It was loaded onto the back of a truck and paraded in downtown Toronto for the public to see. Afterward, the truck with the rocket on top would be seen parked on the street in front of a crowded restaurant as the Da Vinci team continued to party after the official event.

But financing was still the hardest problem for Feeney. Despite the accelerated work to get all the pieces together, he was running out of time. At the rollout he had announced that he still needed about $500,000 for development before he could formally give his sixty-day launch announcement. GoldenPalace.com, the world's largest online gambling site, took up the sponsorship, finally giving Feeney the green light to set 2 October 2004 as his launch date in Kindersley, Saskatchewan. Feeney would himself pilot the maiden flight.

A sense of urgency charged the air, as Feeney determined to outmaneuver the better-financed and better-prepared Rutan, in much the same way that Lindbergh had surprised everyone in 1927 by winning the Ortieg Prize over the favored Richard Byrd. "The x prize purse of $10 million is not a personal inspiration at all," Feeney said in a promotional video. "I'm inspired by the opportunity of physically going into space, of making history. There is no doubt that if we are the first team to go up, we will make history. And that to me is worth a fortune. It's worth a lifetime of achievement.

"You can only put the first civilian into space in a privately built rocket once. It will never happen a second time. It's like bringing the Wright Brothers into Canada."

In the meantime, back at x prize headquarters, both Diamandis and Maryniak were concerned about Feeney. The accelerated pace and motivation to cut corners to beat Rutan's launch attempts were making them uneasy. They sent a team to evaluate Feeney's technical readiness and were prepared to disallow them to fly if they thought the vehicle was not technically sound.

The Da Vinci Project planned a scale model balloon test in Saskatchewan on 12 September. This would eventually be postponed until 23 September to give more time to test key components. Unfortunately, they would not be able to reschedule a launch date. They had run out of time for the prize.

Ken Davidian did not get to visit all of the teams but corresponded with them by phone and e-mail. Interorbital Systems was headed up by married couple Randa and Roderick Milliron. They founded the company in 1996 and started designing and testing rockets at the Mojave Spaceport in California. Their two-stage manned vertical take-off Neptune-Solaris spaceliner was to be their entry in X PRIZE competition. At one point, they had enlisted the support of the King of Tonga, who was eager to use his island's convicts to build a spaceport.

Other international registrants included De Leon Technologies, headed by Pablo de Leon, essentially a one-man space program in Argentina developing a single-stage-to-orbit spaceplane. Maryniak recalled visiting his apartment in Buenos Aires much earlier on, saying Leon had a satellite, a "get away special" on his kitchen table, which De Leon Technologies launched in 2001 onboard NASA's shuttle *Endeavour*. X PRIZE took a few foundation board members to witness the launch.

There was also the Israeli team IL Aerospace Technologies, with a two-stage balloon/rocket design. The team was led by Dov Chartarifsky, whose biggest worry, according to Davidian, was "that if he goes up he'll come down in Palestine." And then there was the Romanian team ARCA, whose demonstrator rocket 2B successfully launched in September 2004, using a composite material monopropellant engine. Some of the teams never got past the paper and PowerPoint stage. In total, twenty-six teams from seven countries officially registered for the prize.

X PRIZE Teams List

1. Advent Launch Services, Advent, single-stage rocket
2. American Astronautics Corporation, The Spirit of Liberty, 1.5-stage rocket

3. ARCA Orizont, single-stage rocket
4. Armadillo Aerospace, Black Armadillo, single-stage rocket
5. Acceleration Engineering, Lucky Seven, single-stage rocket
6. Bristol Space Planes, Ascender, single-stage rocketplane
7. Canadian Arrow, Canadian Arrow, two-stage rocket
8. Da Vinci Project, The Wild Fire, two-stage balloon/rocket
9. De León, Pablo Technologies, Gauchito, single-stage rocket
10. Discraft Corporation, Space Tourist, single-stage rocketplane
11. Flight Exploration, Green Arrow, single-stage rocket
12. Fundamental Technologies Systems, Aurora, single-stage rocketplane
13. HARC Liberator, single-stage rocket
14. IL Aerospace Technologies, Negev 5, two-stage balloon/rocket
15. Interorbital Systems, Solaris X, single-stage rocket
16. Kelly Space & Technology LB-X, two-stage plane/rocket
17. Lone Star Space Access, Cosmos Mariner, single-stage rocketplane
18. Micro-Space, Inc., Crusader X, single-stage rocket
19. PanAero, Inc., SabreRocket, single-stage rocketplane
20. Pioneer Rocketplane, XP, single-stage rocketplane
21. Scaled Composites, LLC White Knight, (carrier) two-stage plane/rocket, SpaceShipOne (rocket)
22. Starchaser Industries Ltd., Starchaser 4, (booster) two-stage rocket, Thunderbird (capsule)
23. Suborbital Corporation, Cosmopolis XXI, two-stage plane/rocket
24. TGV Rockets, Inc., Michelle B, single-stage rocket
25. Vanguard Spacecraft, Eagle, three-stage rocket
26. Whalen Aeronautics, Inc., unknown, unknown

By early 2003, with the X PRIZE competition well under way and generating much media attention, Diamandis began to think about what would come next. Now there was too much momentum and media buzz not to keep the ball rolling and plan for the future. He envisioned an annual event of competing suborbital vehicle teams vying for several prizes, perhaps fifty to a hundred launches occurring over a two-week period. It would be called the X PRIZE Cup, the space equivalent of the America's Cup for sailing or the NASCAR auto circuit races.

The X PRIZE Cup would serve as a catalyst for further development among the Ansari X PRIZE teams and foster innovation in space technology. It would also be a means to further global recognition for the Foundation and generate economic growth at the hosting event site. Diamandis gave a young intern, Ryan Kobrick, the job of figuring out exactly how the X PRIZE Cup would work.

Except for some concept artwork by Pat Rawlings depicting a public festival event much like the Reno Air Races or the Oshkosh air show, Kobrick basically had a clean slate in working out the details of the X PRIZE Cup. "I had free rein to be creative," he recalled. Kobrick would conduct an economic impact study to attract hosts and come up with the rules for the prizes within his four-month internship. The competitions would include five major categories: (1) maximum altitude, (2) launch turnaround time, (3) most number of passengers per vehicle, (4) total passengers that can be carried over a two-week period, and (5) fastest flight time. At the end of his internship, he had the draft document needed to start soliciting proposals from spaceports to host the X PRIZE Cup.

By July 2003 an official request for proposals had drawn bids from the City of Lompoc in California, New Mexico's Office of Space Commercialization, the Oklahoma Space Industry Development Authority, and the Florida Space Authority. A rapid selection process occurred in the fall of 2003, aiming for a decision before the one hundredth anniversary of flight.

On 11 May 2004, with an up-front offer of $10 million in state support, New Mexico was selected as the home venue for the X PRIZE Cup. It would be the start of New Mexico's planned developments for an inland spaceport.

And so, by mid-2004, the stage was set for the future of private spaceflight. All that was left now was for someone to actually win the Ansari X PRIZE.

8. Private Manned Spaceflight Makes History

Don't worry, Mike; it's just an airplane.

Burt Rutan

The X PRIZE launch day, 18 May 1996, began with a celebrity-packed press conference announcing the prize and ended with a gala black-tie event at the St. Louis Science Center, at which the prize garnered its first publicly announced competitor. Amid the VIP crowd, which included NASA administrator Dan Goldin, moonwalker Buzz Aldrin, and some twenty astronauts, as well as two of Charles Lindbergh's grandson's, Morgan and Erik, Elbert L. "Burt" Rutan stood up and became the first to declare himself in the race for the X PRIZE.

Burt Rutan had created headlines a decade earlier for building the *Voyager*, a spindly, twin-engine aircraft that touched down at Edwards Air Force Base on 23 December 1986 after becoming the first airplane to circumnavigate the globe without refueling. In fact, Rutan had built a career pushing the limits of aircraft design and taking on projects that others thought impossible.

His exotic looking VariEze, a lightweight, fiberglass composite, kit-built aircraft, had made him a hero among experimental aircraft fliers. He had established a reputation for innovation in executive aircraft (Beech Aircraft's Starship), unmanned aircraft for the Strategic Defense Initiative (Raptor), and rocket fuselages (McDonnell Douglas DC-X). He had also developed a revolutionary rigid sail that Dennis Connor used on his catamaran, *Stars and Stripes*, to win the America's Cup in 1988. In 1996 Rutan was coming off the development of a radically new aircraft called the Boomerang, an asymmetrical, twin-engine craft with a fuselage-mounted engine and a second engine on only one of the wings.

This all meant that when Burt Rutan stood up at the X PRIZE black-tie event and said he was in the competition, people took notice. If anyone could transform space travel in the way intended by the X PRIZE, maybe this was the man. If anyone could usher in the era of private space travel, maybe it would be Burt Rutan. "You know," he had told the *New York Times* the previous summer, "many people look back and say, 'Gee, this is a really boring time, compared to the 60s, when we went so quickly to the moon from first orbit. But I've got a theory that this is just a kind of gentle pause. There's going to be a renaissance, a super-renaissance, in the next fifteen years."

For nearly half of those fifteen years, as the X PRIZE Foundation struggled to gather financial backers, competitors lined up for their shot at the suborbital prize, and Ken Davidian and Gregg Maryniak zipped around the globe checking out the progress of X PRIZE competitors, Rutan labored in complete secrecy at his Mojave, California–based company, Scaled Composites. No one at the X PRIZE Foundation had a clue what Rutan was building out there in the desert until the grand unveiling of his creation on 18 April 2003.

What motivated several hundred aerospace professionals and assorted VIPs to gather on that date in Mojave, in California's windswept high desert, was to learn precisely what Burt Rutan had been up to. Such was Rutan's reputation that this unveiling event drew the likes of Buzz Aldrin, first space tourist Dennis Tito, adventurer Steve Fossett, Air Force brigadier general Pete Worden, Peter Diamandis, Erik Lindbergh, and considerable media.

Rutan's long-awaited entry into the X PRIZE competition turned out to be not one but two new craft, SpaceShipOne, a rocket-powered, suborbital aircraft, and White Knight, the mother ship that would carry SpaceShipOne aloft under its fuselage and launch it from fifty thousand feet. Once clear of the aircraft, SpaceShipOne would fire its single-rocket engine and soar into a steep, eighty-four-degree climb to an altitude of one hundred kilometers, the widely accepted boundary of space. The craft would then glide back to Earth for a runway landing.

"The program is a lot like the X-15," Rutan explained. "But we had a minor annoyance: we had to build our own B-52," a reference to the bomber that carried the x-15 to drop altitude. Attendees at this rollout were treated to a flight demonstration of Rutan's b-52, the White Knight, and a dramatic unveiling of SpaceShipOne from behind a curtained section of the Scaled Composites hangar.

Both airplanes looked as though they had flown off the pages of a science fiction novel. The lithesome, gull-winged White Knight was powered by twin turbojets with afterburners for high-altitude flight. In appearance, it resembled some of Rutan's previous radical designs, such as the world-girdling *Voyager* and Proteus, the tandem-winged plane built to investigate the use of aircraft as high-altitude telecommunications relays.

In contrast, the pod-shaped experimental rocket ship, SpaceShipOne, looked simple and toylike, with its underbelly painted with blue stars. As on the White Knight, multiple round windows dotted its cabin. Although it was a rocket plane, it was actually registered as a glider, a split personality that fit its dual mission of rocketing to the edge of space and then making an unpowered glide back to Earth.

SpaceShipOne's flight profile would submit its pilot to 3–4 g during boost phase. At maximum altitude, he would experience about three and one-half minutes of zero gravity. On the way down, the craft would enter the atmosphere with a "hands-free reentry" configuration, topping out at about 5 g during the descent. After flying some thirty-five nautical miles downrange, SpaceShipOne would land back at Mojave Airport.

As striking as its appearance was its mix of cutting-edge technology and operational simplicity. SpaceShipOne was constructed of an epoxy and carbon fiber honeycomb. Its hybrid rocket motors burned both solid and liquid fuel, with rubber serving as fuel and nitrous oxide (also known as laughing gas) as oxidizer. Operating the engines was simplicity itself. One switch armed the motor and another fired it. There was no throttle, just a sixty-five-second blast of power until the fuel was consumed. The craft would use stick and rudder controls for subsonic flight, electronic controls for supersonic flight, and gas thrusters for positioning itself in space.

The reentry method was pure genius. The ship was designed to reenter the atmosphere without the pilot having to touch the controls. Instead, the wings would rotate upward—called feathering—slowing the craft with their drag, in a sense turning SpaceShipOne from a bullet into a shuttlecock. This maneuver lessened both the g force and the heat of reentry. At about fifteen miles altitude the wings would revert to a normal aircraft configuration, allowing the craft to finish with an unpowered glider landing.

"There is nothing you will see today that is a mockup," Rutan told those in attendance. "I didn't want to start the program until we knew that it could happen." Rutan was already referring to Scaled Composites as having created the first private manned space program, which was no exaggeration. Aside from the two new craft, Scaled Composites had developed a hybrid rocket propulsion system, a mobile propulsion test facility, a flight simulator, an inertial-nav flight director, a mobile mission control center, all spacecraft systems, a pilot training program, and a complete flight test program.

Claims of a private manned space program drew obvious comparisons to NASA's manned space program. Certainly, NASA could afford to do things on a grander scale; however, the contrasts were less ones of scale than of attitude. Burt Rutan followed two principles totally absent from NASA: private enterprise and one man's personal vision.

Although Burt Rutan had been designing about one original aircraft per year for twenty years, he had long resisted involvement with a space project. In fact, avoiding space projects had been a conscious decision he made early in his career.

Born in 1943 in Portland, Oregon, and raised in Dinuba, California, Rutan grew up designing his own model airplanes and getting hooked on flying with his brother, Dick. He earned his pilot license at sixteen, before he had his driver's license.

Like many young engineers in the 1960s, Burt Rutan was profoundly influenced by the early accomplishments of the U.S. space program. "I graduated from college at a time that you would think would compel me to go into the space program," Rutan would later reflect. "It was 1965, halfway between Yuri Gagarin and our first man landing on the moon."

However, after receiving a BS degree in aeronautical engineering from California Polytechnic State University, Rutan chose to work as a test engineer at Edwards Air Force Base rather than join the crop of young engineers and scientists working on the moon program. "I felt I was so far behind on being able to come in and take a new idea and actually get it out there flying if I focused on spaceflight or manned spaceflight."

Edwards had already earned a reputation as the hotbed of design for such revolutionary aircraft as the x-15, the air-launched plane capable of reaching space on suborbital flights, and the sr-71 Blackbird, the Mach 3 reconnaissance aircraft.

After seven years, Rutan left Edwards to take a position as director of flight testing for Bede Aircraft, working with homebuilt kit aircraft. He would eventually use that experience to form his own company, Rutan Aircraft Factory (RAF). He made his home in the dusty town of Mojave, California, a community of a few thousand residents some ninety miles north of Los Angeles. It had little to recommend it other than its dry climate and its isolation. Scaled Composites was founded there in 1982, when Rutan began building airplanes with handcrafted composite materials.

He spent the next decade reshaping the landscape of homebuilt aircraft with a series of radically new designs, including VariViggin, VariEze, Quickie, Defiant, Long-EZ, Grizzly, Solitaire, and Catbird. RAF's crowning achievement came in 1986 with *Voyager* and its nonstop, around-the-world flight, piloted by Burt's brother, Dick, and Jeana Yeager.

Over the years, as he burnished his reputation with innovation and breakthrough aircraft technology, Burt Rutan also became outspoken in his disdain for the large bureaucracies of NASA and the aerospace corporations. In contrast to those huge institutions, Rutan's mantras were "seat of the pants," "hands-on," "simple designs," "get your hands dirty," "just do it."

Rutan was also stamped with audacity. "I don't care about taking the risk that something won't succeed," he said. "That's the big difference between me and the engineers who work in aerospace. Or the managers of the engineers who work in aerospace. They're absolutely frightened of failure." Rutan's brother, Dick, has been known to voice the philosophy more bluntly: "People that run the companies are freaking idiots. No vision. They've got their head up their ass, and they're all bookkeepers."

By the 1990s Rutan had finally turned his attention to space. Scaled Composites built components for Orbital Science Corporation, which was sending small satellites into orbit with a rocket plane launched from a Lockheed L-1011 airliner. Within a few years Rutan was well on his way to developing his own spaceship. He even had a carrier, the Proteus, already built and flight-tested. His plan was to air launch a capsule from Proteus at thirty-two thousand feet. He initially believed that winged aircraft were ill-suited for spaceflight. Following in the footsteps of both NASA and the Russian Space Agency, he was going to use a single-seat capsule and do away with the need for a winged vehicle. "I was going to do a capsule and launch it from an airplane that did a steep climb and a parachute recovery."

As Rutan tells the story, he woke up one morning and realized, "For God's sake, Burt, you've done about forty airplanes. We've got to do this with an airplane somehow."

With the 1996 announcement of the X PRIZE, Rutan's space plans shifted into high gear. Other contenders for the prize were planning to use conventional, vertical takeoff rockets or rocket balloon combinations. But Rutan's thinking was drawn back to the X-15 program and its use of an air-launched rocket plane. The most daunting problem with this approach was not sending the rocket plane up to one hundred kilometers but getting it back.

During reentry, most spacecraft use atmospheric friction to decrease their speed, resulting in extremely high temperatures. Exotic metals or ceramic tiles help spacecraft endure reentry temperatures. However, Rutan's craft would be built of composite materials, which have a much lower resistance to heat.

The solution hit him in another one of those eureka moments in the middle of the night. What if the spacecraft reentered the atmosphere at its most stable configuration by pivoting part of the wing and the tails to create a high-drag configuration? This would allow a much slower descent and prevent dangerous heat buildup during reentry.

When he started showing around his designs, they did not meet with immediate acceptance. "I thought he'd lost his mind," recalls Mike Melvill, his longtime friend, test pilot, and first employee. But Rutan would not be discouraged. The way he characterize his approach is, "Research is doing something where half of the people think it's impossible and half would think . . . hmmm, maybe that would work. A true creative person has to have confidence in nonsense."

Rutan's spaceship designs began as scribbles on paper napkins. He then built large balsa and mylar models that were dropped from a tower to assess the stability of various wing-tail configurations. All of them descended as stable as feathered shuttlecocks. The final determination of which design would be most stable at supersonic speed would have to be conducted through computer simulation.

The only component missing from his plan was financing. In 1996, after Rutan announced his entry into the X PRIZE competition, he met with ardent aerospace buff Paul Allen, the multibillionaire cofounder of Microsoft. They had previously shared ideas about using high-altitude planes for telecommunications, but this time Allen made the pilgrimage to Mojave to hear about Rutan's suborbital spaceflight projects. Allen was interested, but it was still too early in the process. Rutan was not yet ready to use Allen's resources. However, that changed in the spring of 2000, when they met again. "I think I can do this now," Rutan told Allen. "I'm confident enough in this that if I had the money like you have, I'd put it in this." Allen put out his hand and said, "Let's do it."

Out of this partnership came Mojave Aerospace Ventures, which would invest approximately $25 million to develop the launch system and would own the spaceship's technology. By April 2001 the "Tier One" program, as the secret space project had been named, started official development and began to build its team.

Both Scaled Composites and the X PRIZE Foundation recruited staff from the young and eager. At the X PRIZE office in Santa Monica, Diamandis began to build his staff by enlisting International Space University (ISU) master's students on their three-month spring internships. He was taking "the best of the best," he would tell the eager students. Intern Brooke Owens recalled Diamandis saying, "You're going to live, drink, breathe spaceships. It will be the best summer of your life, and if you're amazing I'll do everything I can to keep you."

A number of students, like Owens and Ryan Kobrick, would populate what they termed "cockroach apartments," four blocks from the beach and four blocks from the X PRIZE office, which by now had upgraded from Diamandis's basement apartment to a one-bedroom unit next door. The office, which housed a long folding table with thirty laptops, constantly buzzed with interns. Diamandis's workforce, barely in their twenties, were put in charge of "special operations," which were buzz words for "given a lot to do and little money with which to do it."

"Peter would come in the morning with stacks of paper and announce, 'I've got projects. Who wants some?'" Owens recalled. "No one had any main responsibility; we all had several fires to put out on a daily basis." It was the best environment for any aspiring space cadet to work in, according to Kobrick. "We had a pretty good time . . . having those young, enthusiastic people working together."

Meanwhile, back in Mojave, Rutan was doing the same thing at Scaled Composites with his cadre of young engineers barely out of college.

According to Matt Stinemetze, Scaled's project engineer, "Burt has this unique knack to pull together the best designers on the planet, and drag them out to the middle of nowhere with nothing to do but design really cool spaceships."

Stinemetze was only twenty-four when he started at Scaled Composites, six years before SpaceShipOne's maiden flight. He was a kid from a small town in Kansas. Growing up, he loved everything about airplanes and, like Rutan, was fascinated with building model airplanes of his own. He graduated in aerospace engineering from Wichita State University in 1998, after almost giving up altogether on a degree because he was disillusioned with school.

"I had a real broad résumé," was how Stinemetze described his professional preparation to build spaceships. "I was a model airplane builder. I'm a really hands-on guy. I've been building composites since I was fourteen years old. Learned it on my own. I'm a machinist. I know how to weld. I didn't spend all my time in school just sitting in the classroom getting book smart." He had also helped a Raytheon engineer named Ray Hooper build an airplane in his garage. That connection opened the door for him at Scaled Composites.

Stinemetze submitted his résumé and was hired the next day. Since then he has done his own hiring at Scaled and has come to understand the type of employee Rutan looks for. "At Scaled, they throw you a program, and you've got to know how to run the program, got to know a little bit of everything. If all you know is how to do aero, it's going to be hard to get in there. It's not that I had the best résumé in the world or the best grades, but I had a lot of variety and a lot of things to offer."

After three years working at Scaled, Stinemetze began hearing rumors of a space project in the works. "There was definitely something going on in the background." One day he just happened to be working in the Research and Development Lab when Rutan sneaked in and asked the shop guys to assemble some foam models of a spaceship. Soon Stinemetze would be tagging along and throwing spaceship models off the Mojave airport tower.

In no time, Stinemetze was appointed lead engineer in charge of Scaled's space program. "Most companies would say, 'Hey, you're just a kid out of college. You're not of sufficient level to be in charge of a spaceship.' Here they say, 'We're going to give you a lot of responsibilities and see how you do.'"

For Stinemetze, this sink-or-swim scenario was a dream job assignment. He was in charge of his own airplane project. Only this time it was a spaceship. He knew how to run a team and get the job done. "I was basically the shop liaison," he said of the experience. He also became known as the "get-it-done guy." And perhaps the best aspect of all for any young engineer—he worked side by side with Rutan, whom he calls "the Boss."

"Get it done, do it this way, quit thinking about it," was how Stinemetze described Rutan's work style. "Burt wants to see the 'fire in the eyes.' He wants guys who are not fazed with not knowing everything but get to it and do things. He would say, 'If you don't try anything new, you're never going to have any breakthroughs.'"

Brian Binnie, one of Scaled's test pilots, also described working in Scaled as a "fluid, informal environment where people have many responsibilities. He [Burt] oversees it all. There's Burt, and then there's everybody else. That's great if you want something done. You got a problem, you got an issue, it's one-stop shopping. You just go see Burt, and if you've got your ducks in a row, he may see things your way, problem solved. Maybe he doesn't, so you go back to doing what you think you can do. What I think has made Scaled an exciting place to work in [is that] we don't encumber ourselves with red tape, processes, and a lot of paper work where it's not needed, where we don't think it adds a lot of value."

Binnie adds, "The talent that is out here has been eye-opening. Burt has surrounded himself with people of outstanding character, second to none. He hires a lot of young people straight out of college who haven't been tainted by a big business mindset, where you get pigeonholed very early. Burt likes young energetic types, free-thinkers [who will] roll up their sleeves even if they're not sure it can necessarily be done." The mandate at Scaled was that if you are an engineer, if you design it, you have to then be able to build it.

By the time work on White Knight and SpaceShipOne was in full swing, Stinemetze had some twenty-five people working for him, a meager number compared to the manpower most aerospace companies or NASA have at their disposal. But whether in organizational size or aircraft design, simplicity was counted as a strength at Scaled Composites.

"Logistics kills you," Stinemetze explains. "People think [SpaceShipOne] is this high-tech thing, but it's not. It's as low tech as we can make it to do the job. It's basically a home-built airplane. That's why it's so successful. It's not burdened with all the bugs."

That same simplicity also extended to other phases of the project. Unlike most traditional aerospace companies, which have a budget and the facilities for wind tunnel testing, costing millions of dollars and years of development, Scaled relied on computer simulation testing to ensure stability for White Knight and SpaceShipOne. That made a hero out of a young engineer named Jim Tighe, Scaled's chief aerodynamicist. Tighe's primary responsibility was to make sure SpaceShipOne and White Knight passed all aerodynamic testing through computer simulation.

Another member of the crop of young engineers, he had studied aerospace engineering and computer science at the University of Colorado. After graduating in 1997, he worked for Boeing before joining Scaled.

"Where is Jim Tighe?" was the most often heard refrain among the Tier One team, because he seemed to be able to fix any problem thrown at him. "He is the most talented aerodynamicist that I've met in my career," Rutan boasted of Tighe. "I sometimes think what I could have accomplished in my career if I had been as smart as him at that age."

It would take almost three years for the Scaled Composites team to design, test, and fly White Knight and SpaceShipOne prior to its attempted suborbital flight. SpaceDev would be selected over Environmental Aeroscience Corp Development to develop the "rocket science" of SpaceShipOne's hybrid rocket motor, using a rubber and nitrous oxide fuel. Experimental flights were done in incremental steps, learning and modifying the spacecrafts as they went along, without the high-tech manufacturing facilities generally thought necessary to build a spaceship.

The development of the Tier One program not only involved Rutan finding a sponsor and working out the technical breakthroughs, it also involved dealing with perhaps the more difficult process of licensing a brand-new space vehicle and making sure he had a suitable spaceport from which to launch. This task fell on the Federal Aviation Administration (FAA) Office of Commercial Space Transportation (AST), under the leadership of Patti Grace Smith.

The subject of the government licensing a commercial, experimental, human-rated space vehicle had never been considered prior to XCOR making inquiries back in 1999. Because of Burt's philosophy of keeping things private — "Don't talk about it until you're ready" — Scaled did not approach AST until 2001. That's when Smith, who headed AST, took on the task of defining how the FAA would oversee commercial space launches. It represented a big step for the agency and the high-water mark of Smith's career.

Smith grew up in Alabama during the court-ordered desegregation of the school systems in the South. She and her sister were among the first African Americans to attend a white school. Her determination to "go places" led her to enroll in college at age fifteen. She credits her father with instilling in her a real "can do" spirit at an early age. "If you think you can do it, go for it," she recalled her father encouraging her.

At twenty-six Smith was already a project manager on her consulting firm's biggest contract with the U.S. Navy. She sat on the board of directors before she turned thirty. Later, while working for the Federal Communications Commission, she was given the chance to work for the FAA as associate administrator for Commercial Space Transportation.

"When I went to my first launch, I knew that I had found what I was looking for. I could not contain myself," Smith recalled. "The opportunity to lead an office where you are laying the groundwork, the policy framework, the infrastructure decisions, the governance, the regulations for a form of transportation that I may not see in my lifetime, but my kids will."

Since approving a launch is considered a significant federal action, compliance with the National Environmental Policy Act takes time to process, typically about eighteen months. Each experimental vehicle is different and is therefore handled on a case-by-case basis, with masses of data exchanged and much testing. "Our license covers the operations of the vehicle, not the vehicle itself," Smith explained. "We look at how the vehicle is designed to operate and whether it operates in that way."

"There were people in the FAA who wanted to say this is just an extension of aviation, and I said no. It has similarities and differences. You can't do a one-to-one comparison," Smith recalled saying. "This is a space launch vehicle, 'aerospace,' with aviation and space." According to her, "We wanted to evolve a commercial launch system license. This is history, a brand-new paradigm of testing!" Coordination was important between interested parties, such as the Environmental Protection Agency, the Office of Science and Technology Policy, and Congress.

Many individuals and agencies wanted this commercial endeavor to succeed, but safety was the FAA's governing priority. The populated areas around the Mojave Spaceport needed to be protected. Smith worked closely with Scaled's Doug Shane to work out the details.

The licensing of a private manned spacecraft had never been done before. They would have to learn from each other in the process. "We knew that there was likely going to be tension between the developer, industry, and the FAA, because the government regulator has a different role to play," she conceded. However, because everyone's goal was to make it a safe and successful launch, they knew they could make it work.

Laboring almost around the clock, the AST team managed to get the vehicle and the spaceport operations licensed in the nick of time for SpaceShipOne's planned maiden space flight. "There was so much dependent on it. The government didn't want to fail that," Smith recalled.

On a flight to Mojave to witness Scaled Composites's first suborbital flight attempt, Smith had the overwhelming feeling that something very extraordinary was about to happen. She sat next to a man with his young son. When she learned they were also going to Mojave, she pretended not to know what was occurring there and asked, "Oh, what's in Mojave?" The man replied, "Don't you know? History!"

His comment reflected the general sense building around this launch that it would mark a quantum leap in public access to space. For the first time in over three decades of aircraft development, Burt Rutan had invited the public and press to watch a test flight of one of his new aircraft. For the first time, a Rutan-built craft was heading to space. "We encourage people to come out and bring their children," he said in the days leading up to launch. "So their children can tell their children that they were there."

Around two o'clock in the morning of 21 June 2004, it looked to Patti Smith as if the whole world had responded to Rutan's invitation. The normally quiet and deserted Mojave Airport buzzed with long caterpillar lines of cars and buses queuing up to get into the massive parking lot. The lot itself bristled with campers of those who had spent the night in anticipation of this event. Hundreds of people lined the runway on folding chairs, dozing restlessly under blankets in order to guarantee themselves a prime viewing spot. An estimated twenty thousand eager viewers had come from around the country and across the oceans to see White Knight and Space-ShipOne make their first attempt at a suborbital flight.

The festivities had been going on all night, like a giant tailgate party, with DJs blaring techno music and some devotees dressed as silver-skinned aliens. People were already calling it a Space Woodstock. Patti Smith walked the flight line, caught up in a jumble of sights and sounds and the gathering excitement. "It was cold. I saw people coming in trucks, trailers, however they could get there from all over. Heard different languages." Smith went in the hangar to watch as Rutan's team worked through the final checks on SpaceShipOne.

Meanwhile, in another section of the hanger, Mike Melvill, Rutan's test pilot for twenty-five years, made his final preparations for the launch. If all unfolded as planned, he would be the first private astronaut in history.

Melvill was not your stereotypical space explorer. A sixty-one-year-old grandfather and a high school dropout, he had started out in the family business, making custom machine dies for cutting cardboard boxes. In his thirties, he became a pilot by necessity to help transport the company product. He single-handedly assembled one of Rutan's VariViggens, a compact, two-seat wood-and-fiberglass airplane that Rutan had named after the Swedish fighter plane, the Saab 37 Viggen. Together with his wife, Sally, Melvill flew it cross-country to Mojave just to show it off. Rutan was so impressed he hired him on the spot as his first employee. The couple never left Mojave, and Melvill has since flown every aircraft that Rutan has designed.

Prior to this day's flight, SpaceShipOne had been flown only four times, encountering a variety of problems. For the first drop test from altitude on 7 August 2003, the one Melvill dubs "the most scary flight," nobody knew how SpaceShipOne would fly or even if it could fly at all. Without wind tunnel testing, they had relied solely on the accuracy of Jim Tighe's computer aerodynamic simulations.

White Knight lifted SpaceShipOne to forty-seven thousand feet, where the release took place. Because the craft has no engines, Melvill had to glide in slow deliberate circles, managing his descent to line up for an airport landing. "Flies like a dream!" he exclaimed on the radio as he glided down to a safe touchdown, much to the delight of a relieved Rutan and his ground crew.

On SpaceShipOne's third powered flight, on 13 May 2004, Melvill continued to climb vertically even when his instrument panels went out on him, a bold move that would eventually earn his front-place spot in Rutan's book for the craft's first suborbital flight. "In some places, that would get a test pilot fired," Rutan declared. "In this case, I thought it was a positive that Mike could hang in there and press on."

On the day of the first suborbital attempt, Melvill climbed into Space-ShipOne at 6:47 a.m., after an emotional send-off from Sally, who pinned a horseshoe brooch to the left sleeve of his flight suit. It was a token she had first given to him when she was sixteen, inscribed with their names and the date they met. When he was safely back on the ground, he would return it to her. That was their good luck ritual.

His fate rested less on charms, however, than on the solid trust between him and Rutan. "All of my flying career with Burt Rutan, ten first flights, in aircraft that have never flown before, I've always had to believe what Burt tells me. I have to believe him 100 percent, because if there's any doubts, I'm going to get out and walk away. He's never let me down. I've never let him down, and I don't intend to." Likewise Rutan, with full confidence in his trusted pilot, said, "Don't worry, Mike; it's just an airplane," before withdrawing his head from the cockpit and sending Melvill off.

With SpaceShipOne slung beneath the fuselage, Brian Binnie taxied White Knight onto the runway with Stinemetze on board as flight engineer. "Gee, there's not many people here," Stinemetze observed. Although there was limited visibility from the cockpit, the two men were not prepared for the sight that confronted them when they turned right and rounded the corner. "The window just literally filled up with thousands and thousands of people." Stinemetze remembers thinking, "Holy smoke! This is as big as I think it is. Much bigger than me, and Mike, and Brian, and everything. That's a memory that's definitely going to live with me forever. All of a sudden you suddenly start to second-guess what's happened in your life and say, 'How the hell did I get here?'"

On the flight line, Patti Gray Smith was watching, equally ecstatic as the gleaming White Knight gathered speed down the runway. "When White Knight took off, I felt so privileged to have been a part of this. My heart was racing. I had a sense of the impact of all this on all those young kids. This is the way for kids to understand and get caught up, to see why math and science are so important. They're looking at it! But it was a whole mix of emotions and possibilities that were converging with this one act that we needed so badly on the commercial side of the industry."

It took about an hour for White Knight to reach the release altitude for SpaceShipOne. "I really hated the ride up there," Melvill recalled. "No one wants to talk to you. They think you need to sit there and concentrate on what you are about to do. I really would have liked someone to distract me and have a conversation about something else, because an hour is a long time to sit there and worry about what's going to happen."

At forty-seven thousand feet mission control confirmed a "go" for SpaceShipOne release. After a short countdown by Binnie, Stinemetze pulled the release lever from the rear passenger seat of White Knight. SpaceShipOne dropped away beneath the plane, but within seconds Melvill fired the rubber-and-nitrous-oxide-fueled rocket motor. It slammed him back in his seat with the force of 3 g and propelled the craft smoothly skyward. Nine to ten seconds after engine start, SpaceShipOne went supersonic.

Then, without warning, SpaceShipOne abruptly yawed ninety degrees to the left. Melvill hit the rudder pedal, quickly pulling the plane back ninety degrees to the right. The craft shuddered, and a series of loud bangs emanated from the back of the ship. He thought he was about to lose control. His hand hovered above the engine cut-off switch. But he rode it out, allowing the full engine burn to blast him out of the atmosphere. SpaceShipOne coasted through the rest of its trajectory, up to 328,491 feet, beating the X PRIZE-required altitude by a fraction. Melvill had time to throw two handfuls of M&Ms from his sleeve pocket and look out the windows before his three and a half minutes of zero g was up. He chose the M&Ms to visually demonstrate zero g, because they would not jam any of the instrumentation when he reemerged from microgravity.

"The sky was jet black above, and it gets very light blue along the horizon," Melvill recalled at the postflight press conference. And the Earth is so beautiful, the colors of the Earth, the colors of the high desert, and along the coastline. It blew me away."

Back at mission control, everyone, including Allen and Rutan, watched the altimeter reading until it barely crept past 328,000 feet, thus meeting the necessary 100-kilometer altitude requirement of the X PRIZE. They had planned to get as high as 360,000, but the problems during the launch had taken a toll on the maximum vertical lift. During the actual X PRIZE qualifying flights, however, they would have to exceed this altitude while carrying an additional four hundred pounds, representing the weight of two passengers.

In preparing for the landing sequence, Melvill tried to move the trim stabilizers or "stabs" from launch "nose up" to centered landing position. They wouldn't budge. He worried that the tremendous bangs he had heard just before engine cut-off might have damaged the tail of the plane. He was not aware at the time, but the bangs had only been pieces of unburned rubber fuel flying out the nozzle. However, because he could not see the tail booms from the cockpit to assess any damage, he could not determine what the problem was. Bottom line: not being able to straighten the stabs meant an uncontrolled descent, which Melvill could not survive.

When he later arrived back on terra firma, Melvill commented that during those harrowing moments, he thought he was going to be a "squashed bug." At one time or another, he had considered aborting the flight or trying a high-risk bailout. "I had a sort of resigned feeling in my mind that there was no way to get back with a situation like that."

Fortunately, after switching to his backup power system, he was finally able to straighten the stabilizers to landing configuration. Still, it was not until the chase planes confirmed that the spaceship was intact that Melvill stopped stressing about the loud bangs he heard and whether they would prevent him from landing safely. Melvill returned to Mojave an astronaut, earning the first civilian astronaut wings from the FAA.

The idea of the astronaut wings did not exist until Michelle Murray from the FAA/AST office asked Patti Grace Smith one day, "You know military astronauts have wings, NASA astronauts have wings; this is a commercial flight—why don't we have wings?" Smith asked if she had any ideas for that. Murray had done her homework and had already researched the design and the equivalent NASA qualifying requirements. Smith took the idea to then FAA administrator Marion Blakey and got the wings concept approved.

At the postflight press conference, Smith was on hand to do the honors by presenting Melvill with his civilian astronaut wings.

The great significance of the event had begun to sink in. For Smith it was "a whole mixture of emotions and possibilities that were converging with this one act that we needed so badly on the commercial side of the industry." She well recalls that moment when she pinned the wings onto Melvill's jacket. "He was just welling up with emotion. I whispered to him, 'Do not cry. Please do not cry.' Because I would have lost it! I was holding together as best as I could. He and I got a good laugh out of that."

Following the flight, Melvill paraded along the rows of people lining the tarmac, standing on top of SpaceShipOne while being pulled by a white pickup truck. At one point Rutan, riding on the truck, came running toward the crowd and returned with a white sign. He gave it to Melvill to show the crowd for the photo-op. The sign captured the moment: "SpaceShipOne — GovernmentZero."

The sign recalled Rutan's long-running criticisms of our government-run space program, a sentiment he had clearly expressed at a press conference just the previous day. "Thirty years ago, if you had asked NASA — and people did in those days — 'How long would it be before I could buy tickets to space?' the answer was, 'About thirty years.' If you ask today, you'll get about the same answer: thirty years. I think that's unfortunate. There has been no progress at all made towards affordable space travel."

In the minds of the throng, cheering as Melvill and Rutan paraded along the runway, there was no doubt that they had just witnessed a monumental leap in that progress. Some in the crowd reveled in a glow of excitement for the role they had played in this achievement. "Oh, my God. This is history; it's being penned right now," Brooke Owens recalled thinking. "I can touch the ink while it's still wet. It's fantastic!" In that instant, Ryan Kobrick and Brooke Owens felt as though they were "going to change the world."

After Melvill barely reached one hundred kilometers, there was a lot of apprehension at Scaled Composites about tackling the X PRIZE weight requirement. They busily worked on boosting their rocket motor and stripping the spaceship of nonessential weight components to carry four hundred pounds more ballast, the equivalent of two more passengers, as required by x PRIZE rules. Matt Stinemetze would end up taking off extra threads from bolts just to shave weight here and there.

Meanwhile, shortly after the launch, Peter Diamandis received a phone call from the company insuring the $10 million prize. They wanted to meet. When Diamandis sat down with the company representatives, he offered to give them at least some benefit from the gamble he thought he was about to win. "You guys are going to lose this thing. Pay us an additional half million and we'll make you a sponsor for the competition," he said. He wasn't prepared for what came next.

The insurance representatives said they had actually invited him up because they didn't think Burt Rutan would be successful when SpaceShipOne had to carry the four hundred pounds of additional weight required for the prize. They suggested that Diamandis lower the altitude rules and pay out only $5 million or extend the period by one more year. They "tried to renegotiate the deal with me," Diamandis recalled dumbfounded. "I just packed up my stuff and left!"

On 29 September 2004 an even larger crowd of tens of thousands, including celebrities and hundreds of reporters, gathered at predawn in the cold Mojave Desert to see the first of the x PRIZE flights. Walking around the cordoned-off VIP grounds, among mockups of some of the x PRIZE contenders dubbed the "Rocket Garden," was nearly every high-stakes player in the private human spaceflight business. The press and camera crews followed Diamandis and Eric Anderson as they made their rounds. x PRIZE contender Chuck Lauer was also on hand to witness the launch.

A subset of attendees included big-money investors and would-be space travelers who had the interest and resources to pursue their dreams. Elon Musk of SpaceX and *Titanic* director James Cameron talked to the press about their plans. Space Adventures brought in a busload of its aspiring suborbital and orbital clients. The x PRIZE interns had their hands full with VIP guests, including forty of Anousheh Ansari's relatives.

This event was demonstrating that space had definitely developed a cache among the wealthy, offering them everything from a $30 million visit to the International Space Station to five-figure deposits to place their names on the waiting list for a suborbital flight. And then there were the high-tech "thrillionaires." Rick Tumlinson had already remarked that the promotion of public access to space had become something of a geeky status symbol. "It's not good enough to have a Gulfstream V, now you've got to have a rocket."

"Space geeks" who had made their fortune in such technology-related ventures as PayPal (Elon Musk), Amazon.com (Jeff Bezos), Google (Larry Page), and computer games (John Carmack) were now directing their wealth into creating vehicles to carry people to space. Peter Diamandis, in acknowledging the rise of space money men as a unique moment in history, declared that "there is sufficient wealth controlled by individuals to start serious space efforts."

On this day, Microsoft multibillionaire Paul Allen's $25 million investment in Burt Rutan's X PRIZE quest was about to face its first full test.

Rutan had originally picked Pete Siebold to pilot this first X PRIZE launch. A test pilot and a brilliant computer engineer, Siebold had designed the SpaceShipOne and White Knight flight simulators from scratch, as well as both Tier One navigation units. He had already piloted two of the four SpaceShipOne rocket test flights. But an enlarged spleen, thought to be cancerous, grounded him. Weeks later, when the cancer scare proved to be a false alarm, it was too late for him to get back into training for the flight. Mike Melvill once again got the assignment.

Gregg Maryniak kept busy on the podium as master of ceremonies. Huge video monitors above the crowd displayed the unfolding events, and live coverage went out over television and the World Wide Web. "It was the biggest webcast in history. We were sandwiched in between Britney Spears and the Victoria's Secret fashion show," Maryniak recalled. When Melvill slipped into SpaceShipOne's cockpit and White Knight lifted off from the runway, video cameras inside and outside SpaceShipOne captured every instant live.

Spectators in Mojave and viewers around the world watched video feed as the flight unfolded smoothly, until the end of the rocket-powered boost. A camera mounted on SpaceShipOne's tail, suddenly revealed a wildly spinning view of the Earth. Something had gone terribly wrong. In the cockpit, Melvill struggled to regain control of the ship. While another pilot would have shut down the engine, Melvill had to let the engine burn if he hoped to reach the required 328,000-foot altitude.

After twenty-nine spins outside the atmosphere, he finally regained control by using compressed-air thrusters from two air bottles that fed through the reaction control system. He managed to take a few photographs out the window at 337,700 feet before heading down to a flawless landing. Melvill went 10,000 feet over the required altitude. While still on the glide path to the runway, Melvill did one more surprise victory roll to round off his total to thirty spins.

"We knew what we had to do. My task was not to damage the airplane. I wasn't going to go for any altitude records but just plenty of margin and burn the engine as little as possible and land the airplane as smooth as possible so we didn't have to fix anything. We didn't even change the tires. We refueled it, and it was ready to go. We could have gone the next day," Melvill said.

On the other hand, Jim Tighe and the rest of the engineers had to go back to the simulator and wrestle with Melvill's control issues. They needed to demonstrate a clean, controlled flight. A few days after, Tighe had it fixed. It was now up to the pilot to deliver in practice.

In between the two SpaceShipOne x prize launches, the x prize staff went back to Los Angeles and Santa Monica for a breather. On the evening of the final launch, Kobrick and Owens were back in the makeshift Mojave x prize office. Maryniak and Angel Panlasigui, then x prize office manager, were preparing all the memorabilia that was to fly as ballast to make up the weight requirement in place of the two other passengers. Kobrick recalled, "When she was getting that ready, she turns to Brooke and myself and said 'Hey, is there anything you guys want to fly, something small?' and I said, 'Yeah, absolutely.'"

Kobrick sent up his iron ring taped to a business card, while Owens did the same with her isu master's graduation ring. Scaled Composites employees also were given the chance to throw in their own personal items for the flight, including wedding rings and models. All the items flew on the winning flight, together with other personal items from numerous people, including a memory stick that contained all the names of the x prize donors who had signed up online and gotten certificates.

Less than a week from the first launch, SpaceShipOne and White Knight rolled out again on 4 October, the forty-seventh anniversary of the historic 1957 launch of *Sputnik*, the event that ignited the U.S.-USSR space race. This time the fate of the prize and the success of SpaceShipOne's seventeenth flight rested on the shoulders of pilot Brian Binnie, who had not piloted the craft since the first rocket-powered flight on 17 December.

The 17 December flight had celebrated another memorable occasion, the hundredth anniversary of the Wright brothers' flight at Kitty Hawk. Binnie had been chosen for that SpaceShipOne flight because he had the most experience with rocket engines and had been the test director for all the rocket firings. "There were a lot of unknowns. We were going to go get this vehicle to fly supersonic on its first rocket flight attempt," Binnie explained. The flight itself was "like drag racing. Vehicles accelerate from zero to three hundred miles an hour in three seconds. It's the same way the spaceship takes a hold of you. But it doesn't stop after three seconds. It keeps going. You have to control the dynamics during the first few seconds. All of these things are happening while you're still trying to catch up with yourself."

On that flight, SpaceShipOne successfully went over Mach 1 and became the first privately built supersonic airplane. Unfortunately for Binnie, it ended with a hard landing, collapsing the craft's left landing gear as it skidded off the runway into the dusty desert floor. Binnie thought he had lost his chance to be an astronaut. "Words cannot describe how disappointed I am," Binnie said to an upbeat Rutan after the landing. "It was just a heartbreaker, no doubt about it. For me it was the TV ABC sports 'the thrill of victory and the agony of defeat.'" He certainly thought he would be at the end of SpaceShipOne's pilot queue or might never get back in the lineup.

Eventually, he had to convince himself that surely he hadn't "come this far to be defined by a landing. The bottom line is I had to try to position myself that if the door opened again and I was given the opportunity, then I'll be ready." And with that he did. "I lived in the simulator. I was sick and tired of flying the simulator, but I think I knew every nuance of the vehicle. I paid my dues that way." He started running to keep fit, which was painful after four knee operations. "As I ran, I would rehearse the entire flight profile. And that sort of became my penance, my discipline for giving me the opportunity that I'd be ready to step up to perform."

So, three days prior to the second launch attempt, Binnie was rewarded with the honor of piloting the aircraft that would attempt to win the X PRIZE. It was a pretty intimidating position to be in. Rutan needed a flawless flight to attract commercial business. They needed the performance to get to altitude but also demonstrate the precision of SpaceShipOne. Binnie thought that choosing him to fly made a better story for the public and for the investors. "If a guy who has not even flown the vehicle for ten months can get up there and win this prize, then you should have a lot of confidence that this would be a great product to invest in." On the other hand, he could not even contemplate the repercussion of a bad performance. "To me [it] was an abyss, just a bottomless pit. The downside was unimaginable."

Binnie had come a long way for his shot at astronaut wings. He was a fifty-one-year-old veteran Navy test pilot, who had also flown more than thirty combat missions during the Gulf War. Born in West Lafayette, Indiana, he moved to Scotland at age five with his family and then returned to Boston as a teenager. Inspired by the Apollo era and his mother's own desire to be an astronaut, he tried to steer his career in a way that kept the possibility for spaceflight open. He was not thrilled with the prospects of sitting behind a desk or flying commercial airlines when he left the Navy. On a tip from a friend, Binnie traveled to Mojave, California, to look into job with a new start-up company called Rotary Rocket that was building its own spaceship. Gary Hudson hired him as a test pilot, which turned out to be a less than enviable position. The Roton was "barely flyable," Binnie recalled. "Sitting on top of a tank of 4,000 pounds of hydrogen peroxide that would melt you." It was not a desirable place to be.

When asked where he found the self-confidence to undertake such dangerous assignments, his response turned first to his past participation in sports, which he termed a "measured form of aggression and physical confidence." As for the actual flight testing of experimental aircraft: "I really think the key ingredient to confidently fly a vehicle is to be a participant in its development—initial design, manufacturing, flight test. You really understand the compromises, sensitivities, strengths and weaknesses of the vehicle design and structure. That knowledge gives you a lot more insight in the way you design a flight test effort. You foresee which parts are dangerous parts of the flight."

Binnie worked for Rotary Rocket in Mojave until the company went bankrupt, but by then he had already lined up something better. It just so happened that golf, one of his passions, was also Rutan's sport. The two met and played together on the golf course for two years. When Rotary closed down, Rutan said, "We'll find something for you to do." Binnie started working for Scaled Composites in 2000, just about the time the financing was arranged with Paul Allen.

The night before Scaled Composites's second x PRIZE attempt, Binnie thought hard about the next day's flight while trying to sleep on the living room couch with his dog. He had graciously given up the master bedroom to his in-laws. Tossing and turning through the night, he couldn't get out of his head just how very much was riding on his performance tomorrow. The flight would not only win or lose the $10 million x PRIZE but seal Rutan's deal with Richard Branson to create a commercial spaceline to take passengers to space.

As if the stakes were not high enough, he heard Rutan boast to anchorman Miles O'Brien on national TV the previous evening that he "not only thinks they're going to hit a home run; it's going to be a grand slam." Binnie slept only about two and a half hours that night and arrived at the Scaled Composites hangar early for his 4:30 a.m. preflight briefing.

However, all of the scenarios he rehearsed in his head, all of his meticulous preparation, did not prepare him for one surprise that unfolded prelaunch. After suit-up and flight briefings, Binnie was on his way to the aircraft in the early dawn to make his own bit of history when his mother-in-law stepped out of the crowd to give him a hug, carrying a large cup of McDonald's coffee. As they hugged, the coffee spilled down his flight suit. There was no time for a suit change, and so he entered the cockpit smelling like Vanilla Roast.

One of the engineers who's made all the calculations on the performance of the vehicle pokes his head in and says if I'm wearing twelve ounces of a sixteen ounce cup of coffee, I'm effectively wearing 400 feet of apogee, or height, and 400 feet was the margin by which Mike Melvill made it across the 100 kilometer line in his very first try to get there. On that note, he closed up the cabin door and I have that thought sort of riding with me on the hour flight up to the release point. It was hard work the whole way. It took a lot of focus to stay on the bright side, stay relaxed enough to execute what we thought needed to be done.

At 47,100 feet, Stinemetze released SpaceShipOne again, and an instant later Binnie fired the rocket, passing so near the carrier plane that he filled the White Knight cockpit with the roar of the hybrid rocket. "Holy crap, that was close," Stinemetze exclaimed. Binnie climbed vertically with no problems.

One minute into the flight, SpaceShipOne exceeded three times the speed of sound, a record for any civilian aircraft. "I went scooting right through the X PRIZE altitude and past the X-15 old record by 13,000 feet or so. I got to the point after rocket motor shutdown and the 'feather' coming up, and I hadn't touched any of the reaction control system yet to control body rates. The vehicle was just absolutely stable. I actually used reaction control to give myself a different view so I could take some pictures."

Binnie experienced more than 3.5 minutes of zero g and flew a little model SpaceShipOne in the cockpit. He did not release M&Ms in space, but the running joke was that Doug Shane in mission control could hear faint cracking sounds through the radio during the one-hour climb, as if Binnie were eating his M&M stash while waiting for release.

The flight itself was an amazing experience for Binnie:

It's such an exhilarating ride that gets you there. It's not a passive experience. It really gets the heart rate going. Noise and vibrations, accelerations of the motor that last about a minute and a half. There's such a sharp contrast when the motor shuts down because three wonderful things happen that I can refer to as the "Holy Trinity" of spaceflight. The noise goes away, the shaking and the vibrations, and the motor goes away, and you cross this line into instant weightlessness. Weightlessness is such a wonderful feeling.

And then on top of that you have the opportunity for the first time to see the view, this view that you've never seen before. And everything that you are seeing with your eyes—just wow! Look at that. There's San Francisco, there's Baja Mexico, there's the Pacific, there's the Sierra Nevada Mountains. There's weather patterns I've only seen on the evening news, and you couple that with this tension-free feeling of weightlessness.

So everything you're feeling is "Wow. This is so cool!" I don't think anybody can have that experience and be depressed. I felt like I had gone through the wringer in terms of the emotional tension, the anxieties with the crash, and the waiting period, the simulations and rehearsals, and the drama of the last few days, and now here all of a sudden I'm being rewarded of this childlike wonder and view of the Earth below. Anyway, I think it's an experience that we won't have to try very hard to sell, because it will sell itself.

Binnie returned to Mojave with a picture-perfect landing and paraded around the block on top of SpaceShipOne carrying a U.S. flag. After eight years, the prize was finally won. It proved that a small company with a handful of brilliant engineers could indeed run a space program for about 2 percent of the cost of a single shuttle launch. SpaceShipOne, having flown again within a matter of days, had proven the concept of reusability, a feat no other space program or nation could claim.

Rutan put things into perspective. "Anybody who goes out and tries to do manned spaceflight has a different benchmark to compare themselves against. It's no longer insane to think of a commercial company doing a manned space flight."

No longer insane, indeed. Halfway through the flight, when Space-ShipOne's stabilizers had locked in place for a safe return, Richard Branson shook Paul Allen's hand and slapped Burt Rutan on the back. The gestures confirmed the handoff from the spaceship development program, financed by Allen, to the commercial spaceline project, Virgin Galactic, to be created by Branson. The age of commercial spaceflight was poised to begin.

On the one-year anniversary of winning the X PRIZE flight, SpaceShipOne was unveiled in the National Air and Space Museum in Washington DC. It was hung from the ceiling of the Milestones of Flight gallery to a packed crowd of visitors and reporters, among the Bell X-1 (flown by Chuck Yeager) that first shattered the sound barrier, the *Spirit of St. Louis*, and the Wright flyer.

Attending the event were Allen and Rutan. Rutan had chosen to display the spaceship in its original state after the 21 June maiden suborbital flight, stripped of its Virgin Galactic and sponsorship logos, with a crumpled fuselage showing the technical difficulties Melvill had encountered on the historic flight. The markings on the plane and the display information below simply stated "A Paul G. Allen Project." It seemed a little out of place in the museum full of relics from the past because it represented the future and the beginning of what was yet to come. For anyone who had seen it taxiing off in Mojave and returning to a thunderous, cheering crowd of thousands after three successful launches, it seemed anticlimactic to see it retired so soon.

As the ceremonies went on, Rutan kept looking up at his craft from the side podium, seemingly mesmerized and oblivious to the speeches or to the hundreds of people around him. For the moment, it was just him and the spaceship he lovingly built. SpaceShipOne was already enshrined as an icon of spaceflight history before it had fully played its role in defining spaceflight's future.

9. Space Tourism Goes Mainstream

Screw it; let's do it.

Richard Branson

In September 2004, as the days wound down to SpaceShipOne's first X PRIZE attempt, the age of commercial manned space flight got its own official launch halfway around the globe, in London. Sir Richard Branson, billionaire owner of the Virgin Group, stood beside Burt Rutan to announce that he was starting a new business, a spaceline named Virgin Galactic that would offer suborbital spaceflights for $190,000 per ticket.

It was a bold plan from a man who had built a career out of making bold plans profitable. Business analysts had concluded that space tourism could develop into a lucrative niche market. Few could afford the $20 million to fly to the International Space Station, but at $190,000, it was predicted, people would stand in line to go into space.

Branson had put a down payment on his space business by striking a $21.5 million deal with Rutan and Paul Allen, who owned the rights to Space-ShipOne. Virgin would invest a further $100 million to purchase five, second-generation versions of SpaceShipOne (to be called SpaceShipTwo) that would be carried aloft by two second-generation White Knight mother craft (to be named WhiteKnightTwo). Additionally, Virgin would build the ground facilities needed to operate the fleet.

Virgin would begin taking deposits the following year, Branson explained, and within five years planned to fly three thousand new astronauts. Countries could fly their own astronauts, while individuals could fulfill their dream of viewing the majestic beauty of our planet from space. Over time the price of flights would decline, putting it within reach of tens of thousands more. If it was a success, Branson planned to move into orbital flights and possibly even an orbiting space hotel.

For their money, Virgin Galactic flyers would get a three-day space experience: medical tests, a trip debriefing, and time in a simulator on day one; a White Knight flight on day two; then their trip to space on day three — not to mention deluxe accommodations and meals.

Five years earlier, Branson had become Sir Richard when he was knighted by the Queen of England for "services to entrepreneurship." It was a fitting tribute for the renegade capitalist who had become a cultural hero for shaking up the stuffy world of big business. Since the 1970s, Branson had hit the United Kingdom and subsequently the entire world like a free enterprise tsunami, audaciously launching his trademark Virgin Group into industry after industry. Virgin Records came into being in the early 1970s, followed shortly by Virgin Broadcasting and Virgin Communications. By the 1980s Virgin was plunging into several different businesses a year. Hotels, publishing, cosmetics, cars, retailing, airlines, energy — the Virgin brand went everywhere. Now Virgin would be mated with Burt Rutan, who had created the first commercial aircraft to travel into space.

In the coming months, the world gained an inkling of how Sir Richard would tackle the challenge of commercializing space. He would leave the tricky business of actually getting to space to Rutan. By 4 October Rutan had won the X PRIZE and had a second-generation, passenger-carrying spaceship on the drawing boards. Branson's flair would initially focus on the marketing.

In the fall of that year, Branson was also knee-deep in the launch of yet another project, a TV program in which he would star called *Rebel Billionaire*, scheduled to premier on 9 November. It was modeled on the NBC reality program *The Apprentice*, starring another larger-than-life businessman, Donald Trump. The show would introduce the rogue Branson to the American public as something of an anti-Trump. Branson was an advocate for fun in the workplace, not intimidation. Eager Virgin interns on the show would not be sitting behind a desk creating spreadsheets, á la *The Apprentice*, but skydiving or engaging in other adventurous challenges, just like their adventure-seeking mentor. Branson had earned his success not by following the stodgy rules of business but by breaking them.

In Hollywood, New York, and on the TV talk shows—everywhere he went to promote the show—Branson boldly mentioned he would be taking people to space. It brought mega-media exposure and an uncharacteristic glitter to the private space arena, as well as putting on public display Branson the showman.

On *The Howard Stern Show*, Branson revealed just how different he was from a run-of-the-mill space entrepreneur by explaining what first spurred his interest in space. Unlike so many of the people working in the space business, whose interest was inspired by *Sputnik*, von Braun, or the Apollo flights, Branson claimed his own fascination sprang from becoming painfully aroused while watching Jane Fonda in the 1968 movie *Barbarella*.

Whoa! This was definitely not the visionary Wernher von Braun talking about Mars missions on the Walt Disney show or the cerebral Gerard O'Neill calculating the physics of space colonies. It wasn't the reclusive Burt Rutan explaining the lift-to-drag ratio of his new craft or some techno-billionaire standing quietly in the background while financing his dream project. This was Branson at his masterful best, having fun and focusing public attention on his projects. He'd once dressed in full drag in a frilly wedding dress to launch his Virgin Bride shop.

Branson had often been criticized for such eccentric behavior and for diluting the Virgin brand by connecting it with too many different ventures. However, all the promotional gimmicks aside, Branson's decision to step into any new venture was always based much more on careful calculation. "The time to go into a business is when it's abysmally run by other people," he has said, "or when Virgin can offer a significantly better customer experience."

In the fall of 2004, Sir Richard was putting the world on notice that he would be mixing high finance and business acumen with his own flamboyance to take the industry of space travel in a whole new direction.

Richard Branson's first connection to Burt Rutan resulted not from a business venture but from his own involvement with epic adventures. Branson has said that he believes in "living life to the fullest," and whether in business or adventure, he has done just that. Branson's entrance into the entertainment and travel fields inspired him to begin a string of daredevil adventures that have now branded his character.

In 1984 he joined a team of sailors on the catamaran *Virgin Atlantic Challenger* to capture the record for the fastest transatlantic crossing. They were forced to ditch sixty miles from the coast of Britain when the boat's hull split open. Undaunted, he returned the following year and succeeded in beating the record by two hours.

After logging that maritime achievement, he took to the air for a series of ballooning quests. He flew with Swedish balloonist Per Lindstrand across the Atlantic, again ditching at sea and almost costing both their lives in the process. Their subsequent two-man flight across the Pacific from Japan to northern Canada in the massive hot air balloon *Virgin Pacific Flyer* ended up setting another world record in January 1991 as the longest flight in lighter-than-air history. In 1998 Branson, Lindstrand, and renowned aviator Steve Fossett set out to be the first to fly a balloon around the world. After launching from Morocco, the three men flew for seven days and covered more than sixteen thousand miles before the attempt ended with a harrowing escape from death when they crashed in the Pacific Ocean, fortunately near rescue services in Hawaii. Descending toward the ocean and fearful that it could be the end, Branson had hurriedly scrawled instructions to his family where he wished to be buried if they were to retrieve his body.

Branson was unable to make another attempt at the record before the team of Bertrand Piccard and Brian Jones in the *Breitling Orbiter* successfully circumnavigated the globe in March 1999, putting an end to the coveted first round-the-world balloon race.

At a loss for new adventures of his own, Branson financed a quest for Steve Fossett to fly an airplane nonstop around the world. It was this endeavor that first brought Branson into a working relationship with Burt Rutan, when Scaled Composites was commissioned to build the Virgin Atlantic *GlobalFlyer*. Somewhat resembling the famed wartime P-38, with twin tail booms outboard of a smaller, central nacelle, the spindly jet aircraft had a solitary turbofan engine mounted atop the manned central fuselage, behind the pressurized cockpit. Piloted by Fossett, the aircraft would eventually achieve an aviation milestone in 2005, completing a solo, unrefueled circumnavigation of the globe in the record time of sixty-seven hours and one minute.

In 2003 Will Whitehorn, then Branson's project manager, flew to Mojave on *GlobalFlyer* business. While there he witnessed SpaceShipOne and White

Knight being built on Scaled Composites's shop floor. Whitehorn immediately phoned Branson, his voice full of excitement and a few well-chosen expletives. "Forget the *GlobalFlyer*," he urged. "He's building a spaceship!"

The news could not have fallen on a more receptive audience than Richard Branson. Influenced by his mother's aviation passion (she had been an aviatrix during the 1940s) and his own fascination with the moon landings, Branson had always believed that someday he would get the chance to fly in space. In 1979 he made one of the United Kingdom's most successful semi-documentaries, *The Space Movie*, in honor of the tenth anniversary of the first manned moon landing. When glasnost came about, Branson was one of the first to contact the Russians to see if he could go up in one of their Soyuz spacecraft. Nothing came of it, and in 1991 the world witnessed the dramatic collapse of the Soviet Union.

In explaining Branson's evolving interest in space, Whitehorn recalls an incident from 1995, when Branson, Whitehorn, and moonwalker Buzz Aldrin were sitting in a bar in Morocco, waiting for better weather to launch yet another round-the-world balloon attempt. Branson wondered aloud why rockets had to be launched from the ground instead of being air launched from something like a balloon. Aldrin explained that they could indeed be launched that way, mentioning projects such as the x-15, which launched from a carrier aircraft.

Whitehorn remembers Branson telling him to register the Virgin Brand and keep an eye on space. And yet, at the time, the commercial potential of space was far from obvious. In 1998 Branson had turned down Peter Diamandis's request of support for the x PRIZE competition. As Branson explained in his autobiography, space flight still "seemed like an impossible dream."

However, as the x PRIZE competition built momentum and garnered worldwide interest, Whitehorn eventually registered the Virgin Galactic name and started looking into the teams vying for the prize. So, when Whitehorn came upon a spaceship in Scaled Composites's hangar in 2003 and reported it to his boss, Branson immediately called Rutan and asked, "Are you really building a spaceship?" Rutan acknowledged that he was, but that he could not disclose whom it was for.

Whitehorn had already visited Rotary Rocket—another x PRIZE competitor in Mojave—but had decided that their use of a helicopter recovery system would not work. However, the more the Virgin team learned about Rutan's efforts, the more they were convinced he was on the right track. SpaceShipOne's design "ticked all the right boxes for somebody like Sir Richard and myself," Whitehorn noted. They liked Rutan's approach of thinking through the safety issues for space, and thought that an air-launched approach was the ideal way for space tourism and for all the other possible uses.

Burt Rutan would later pitch Branson's interest in the project to his financial backer Paul Allen, telling him that somebody wanted to commercialize the spaceship if it was successful. Delicate negotiation between the billionaires eventually resulted in a deal between Virgin and Mojave Space Ventures. Virgin would purchase an exclusive license to SpaceShipOne's core design and technologies and would order five passenger spaceships from Scaled Composites for $50 million. A roughly equivalent amount would be invested in operations and infrastructure, bringing the total above $120 million.

Even for the go-for-broke Branson, it was a considerable gamble. The aircraft that would fill his space fleet did not yet exist. And when it did, one single accident might completely destroy the market for private space flights. A regulatory morass also had the potential to delay or even scuttle the project.

However, in the grander scheme of things, two overriding considerations outweighed the risk. For one, the investment was manageable. The previous summer, Virgin Airways had ordered twenty-six new A340-600 Airbus aircraft for several billion dollars to upgrade and expand its fleet. On the other hand, Virgin's total investment to launch an entire space airline—an entire industry of civilian spaceflight—would amount to roughly half the cost of a single Airbus.

Plus, Virgin would get more than just a space airline for its investment. Whitehorn acknowledged that Virgin had been looking for a "flagship company for the 21st century, especially for the U.S. Galactic will put the Virgin brand on the American map in a way that money can't buy."

The second consideration, one that pervades all of Virgin's business projects, is complete faith in the Branson magic and his incomparable ability to generate excitement around a project. He told a *Wired* magazine reporter, "If we can make space fun, the rest will follow." That was the problem. For more than three decades, private space had been the province of space colony dreamers, disgruntled engineers grumbling about NASA, and scientists who grew microgravity crystals in orbit. Where was the fun in that?

Branson, on the other hand, was going to splash the Virgin logo and the Virgin aura onto SpaceShipTwo, onto the whole stodgy business of space travel. While his passengers awaited their flights, he would pamper them with insider briefings, on-location space events, designer space suits, and centrifuge rides, and let them hang out with other A-list space flyers-in-waiting. In addition to their actual trip to space, their three-day launch experience would include medical tests, time in a simulator, a White Knight flight, plus deluxe accommodations and meals. Included in the bargain, Branson would deliver membership into one of the most select, elite clubs imaginable—those who have traveled to space.

Will Whitehorn had experience taking the lead on Branson's cutting-edge technology projects, including *GlobalFlyer* and Virgin Trains. He was given the task of finding financing and convincing the investment committee of the Virgin Group to back Virgin Galactic. The committee challenged him to first find $10 million in deposits to prove there was a market. That requirement was met within nine months, as clients lined up for suborbital flights on the next generation of Rutan's spacecraft, SpaceShipTwo.

Richard Branson was not the first person to hit on the concept of commercializing spaceflights for the masses—and making them fun. In 1968, swept up in the Apollo mania and the release of Stanley Kubrick's *2001: A Space Odyssey*, Pan American Airlines president Juan Trippe announced that Pan Am would start taking reservations for flights to the moon. Pan Am competitor Trans World Airlines quickly followed suit.

Pan Am distributed "First Moon Flights Club" membership cards at no cost, which gave the bearer the opportunity to buy a ticket to the moon as soon as the opportunity became available. By the time Pan Am closed the list in 1971, an astonishing ninety-three thousand people had signed up for a lunar flight, including such notables as Ronald Reagan. The price of a moon ticket was estimated to be $28,000. Wernher von Braun and then NASA administrator Thomas Paine expressed optimism that regular flights to our nearest celestial neighbor would be the norm by 2000 and that ticket prices would come down.

Unfortunately, Pan Am never took any concrete steps toward developing space tourism. The airline filed for bankruptcy in 1991, thereby reducing the moon flight membership cards to historical souvenirs. It would take more than a decade and a different era of entrepreneurs to make the idea of space travel amount to more than just a waiting list.

In the early 1980s, Seattle-based tour operator Society Expeditions, Inc. (SE) was in the business of sending its well-heeled clientele on adventures to the far ends of the Earth — to the poles, the desert, the Amazon rain forest. SE founder T. C. Swartz decided that he wanted to offer the ultimate adventure of all — a trip into space. Swartz approached NASA with a plan to rent a shuttle orbiter and refit its payload bay to accommodate seventy-four passengers, who would each pay $50,000 for a ticket. NASA balked at this potential commercial utilization of the tax-funded transportation system, leaving Swartz with no alternative but to look into the private sector for someone to build him a spaceship.

Gary Hudson's company Pacific American Launch Systems responded to the opportunity by agreeing to lease its planned Phoenix Single Stage to Orbit Vehicle (SSTO) to SE. Financing for the development of Phoenix was not yet in place when Swartz started promoting Project Space Voyage and taking $5,000 deposits for a $50,000 spaceflight seat. He proved the market was there when he accumulated five hundred seat reservations within just one year. The money was held in an escrow account. However, in January 1986 the space shuttle *Challenger* and its crew were lost in an explosion during launch ascent, as a result of which the enthusiasm and market for spaceflight fizzled out. Swartz ended up returning deposits to would-be passengers.

Another decade would pass until the same scheme was resurrected by a few entrepreneurial companies, some of which were vying for the X PRIZE. Jim Akkerman of Advent Launch Vehicles had a similar idea when he established the Civilian Astronaut Corps in 1997. Advent would fund its vehicle development through a $2,000 membership fund that would go toward the price of a suborbital ticket. Harry Dace, Akkerman's partner, started advertising for a 4 July 1999 spaceflight on a yet-to-be-built spacecraft called Mayflower. Within six months of the proposed liftoff, they had fifty-three paid passengers on the list. However, like Pan Am and TWA, the enthusiasm they had built up was far ahead of the vehicle development. They ended up returning the deposits prior to the promised launch date.

Also in 1997, Pioneer Rocketplane began examining options for developing a human-rated version of their supersonic spaceplane design. The vehicle, called Pathfinder, would be used for private spaceflight. "At that time nobody believed in tourism at all. I always thought that it was the growth market," recalled Chuck Lauer, vice president of business development for Rocketplane. Pioneer announced it would launch commercial flights by 2000.

In the tourism sector, two companies, Zegrahm Expeditions and Space Adventures (SA), commenced marketing suborbital flights in 1997, using their experience selling and marketing in the adventure travel business, just as SE had done more than a decade earlier.

Zegrahm Expeditions is a luxury adventure travel company, founded in 1990, that offers high-end soft adventure programs on seven continents. Years of experience and luxury service had endeared the company to a loyal group of high-net-worth clients. Its president, Werner Zehnder, and vice president, Scott Fitzsimmons, had both worked for SE in the 1980s.

Zegrahm established its own space division, Zegrahm Space Voyages, and began offering its suborbital space flight program in 1997 by collecting $9,000 dollar deposits ($5,000 deposit + $4,000 flight insurance) for a $98,000 suborbital ride. The program materials offered a two-and-a-half-hour flight on the Space Cruiser System, to be developed by Vela Technology Development, Inc., an aerospace company headed by Pat Kelley, who had more than thirty years' experience in the space business, including missile warning satellites, launch vehicles, and missile defense projects. The information brochure boasted of Vela's expertise coming from former top aerospace experts within NASA. Vela's two-stage system consisted of a first-stage jet-powered Sky Lifter, which would carry aloft the Space Cruiser, a six-passenger spacecraft capable of traveling to an altitude of one hundred kilometers.

Zegrahm realized that it had had to market not only the spaceflight but everything the experience embraced. The company's sleek brochures outlined in detail what a candidate might expect, including a five-day training program and diagrams of passenger spacesuits, right down to the folding interior seats and flight configuration of the spaceship. Suborbital clients received flight paraphernalia, a flight booking number corresponding to their place in line for a launch, and a suborbital flight ticket as proof of purchase. A list of notable space luminaries as advisers lent authenticity to Zegrahm's offer. By the time Zegrahm sold its space division to SA in 1999, the company had over $1 million in suborbital flight deposits, but there was still no hardware in sight.

With a program similar to Zegrahm's, SA started selling suborbital flight reservations from its founding in 1997, at $98,000 per flight. However, unlike Zegrahm, it had no specific timeline for first launch, which Zegrahm had arbitrarily set for Saturday, 1 December 2001. SA also did not pinpoint a specific vehicle or spacecraft developer for the program. Instead, the company had a menu of at least five developers on its 1999 program catalog, including Rotary Rocket's Roton, Scaled Composites's Proteus, Pioneer Rocketplane's Pathfinder, Bristol Spaceplanes's Ascender, and Kelley Space and Technology's Astroliner. The theory was to use whichever outfitters would deliver the first commercial vehicle. If several developers eventually became flight ready, SA would let each client choose the vehicle on which he or she wanted to fly.

In 1999 Zegrahm Expeditions sold its Space Voyages division to its sole competitor, Space Adventures, Ltd. By the end of 1999, SA was the only company taking reservations for suborbital space flights.

John Moltzan most definitely wanted to reserve a ticket to space. He was barely out of college when he read an article in *Wired* magazine about Zegrahm's Space Voyages being bought out by SA. It was late 1999, and he had just finished a degree in journalism from the University of Maryland. A native German who had moved to the United States six years earlier, Moltzan was an aviation, space, and travel enthusiast and enamored of the prospect of commercial space travel. But since $98,000 was outside his price range, he thought the next best thing would be to work in this cutting-edge business of space tourism.

Moltzan successfully interviewed for a job at the Alexandria, Virginia, office of SA in early 2000 and had soon worked his way up from marketing and public relations to coordinating and growing the number of SA resellers in the United States and around the world. Aside from offering high-end flights to the International Space Station, SA built its business around such related activities as flights on Russian MiG fighter jets, zero-gravity flights, cosmonaut training programs in Moscow, and eventually suborbital flights.

As the business grew, Moltzan transitioned to sales, selling individual and corporate space and flight experiences, a role that included shepherding the company's suborbital clients. This contact with dozens of potential candidates gave him an understanding of the type of individual attracted to suborbital flights. "Everybody had a passion for space for slightly different reasons," Moltzan recalls. They were mostly male, in their mid-thirties and older. There were thrill seekers, pilots, and wanna-be astronauts who had significantly high net worth. Some were lured by the prospect of becoming the first citizen from their country to fly in space.

By 2004 Moltzan was serving as director of business development. Traveling to Asia, the Middle East, Europe, and South America, he managed relationships with international organizations such as the Japanese advertising agency Dentsu and the Japanese Aerospace Exploration Agency (JAXA). He also visited companies that were interested in offering SA's suborbital program and space-related activities as unique employee incentives or promotional products that instantly generated "buzz" for their national or regional marketing campaigns.

Companies such as American Express, Gillette, US Airways, Pepsi, and Volkswagen began to jazz up their marketing with space camps, flights in MiG fighters, and tickets for suborbital flights. "Getting them interested enough to do this [promotion] was not really difficult," Moltzan found. "A phone call or initial meeting often hooked them on the idea, and in some cases these companies would reach out to us based on press coverage they had seen."

However, negotiating with them and easing their concerns could take anywhere from two to six months. They wanted to know when the flight would occur, the safety and insurance issues, refund policies and costs. "Generally, the larger the company, the larger their legal team would be, and the longer the contract negotiations would take," Moltzan observed. Nonetheless, he facilitated SA programs for more than twenty national and regional promotional campaigns in the United States, Europe, the Middle East, Asia, South America, South Africa, and Australia.

According to Moltzan, sales were evenly split between the United States and the rest of the world, and with a few notable exceptions, such as Germany, SA sold suborbital flights in most major markets around the world. SA also implemented its own Spaceflight Club for those who didn't quite have the funds to put down the initial $12,000 deposit. An annual membership of $980 would be applied toward a suborbital flight ticket, and the space enthusiast could attend SA's exclusive events for suborbital clients and be "in the know" about the space industry's progress toward commercial flights.

On the other side of the Atlantic, Moltzan's exclusive United Kingdom reseller, Wild Wings, was busily ramping up business in Europe. Wild Wings was a travel agency headed by John Brodie-Good, an avid bird- and whale-watcher. The company was a wholesaler for all things adventure and had marketed SA's products since the late 1990s. Their Arctic and Antarctic polar programs had given them opportunities to work and deal with the Russians, since their vessels were the pioneers for commercial exploration of the Poles and deep-sea dives. It is this relationship that also opened doors to selling air- and space-related programs in Russia. Up until 2002, Wild Wing's space programs manager was Ian Collier.

Ian Collier had wanted to work in the adventure travel market after traveling around the world and gaining considerable experience in the travel business. Having a bit of an adventurous streak himself, he began working for Brodie-Good at the Wild Wings travel agency, sending clients to Antarctica and the North Pole, to the bottom of the ocean to view the *Titanic* and on supersonic jet flights in Russia. When asked why Wild Wings homed in on selling space-related products, he pointed out that the company had been taking clients to the farthest extremes of the world and to the bottom of the Earth. "Looking up was a natural progression."

Just like Moltzan at SA, Wild Wings's marketing campaign included Web site exposure and direct media attention through press releases and radio and TV interviews. But most of all, Wild Wings relied on a network of third-party resellers doing their own marketing to find individual and corporate clients. "You really do need an effective network of resellers that have contacts in the market," Collier noted. "And just like SA, private individuals have a habit of finding us."

He remembers dealing with the handful of suborbital clients who had booked with SA. "They were pretty relaxed and were there for the long haul." They were mostly private individuals who had the personal ambition to go into space and "were not in it for publicity spin-offs." None of them was too concerned about how long the wait would be and when the flights would actually happen. This was a good attitude to have, since most of these early suborbital clients had to wait years before they saw some tangible progress in the form of Rutan's X PRIZE suborbital flights in 2004. A few of the SA clients on the list were well above age sixty and were also concerned they might be too old by the time space tourism actually got off the ground.

In 2002 Collier left the United Kingdom behind and emigrated to New Zealand. He was working for House of Travel near Christchurch when Virgin Galactic unveiled its worldwide invitation in 2006 for accredited space-flight agents, or ASAS, as part of a global network of travel agents to sell their suborbital flights. Today, ASAS are the only travel agents allowed to reserve Virgin Galactic suborbital spaceflights. House of Travel, the largest independently owned travel company in New Zealand, was appointed Virgin Galactic's exclusive agent in that country.

Collier's experience in selling space-related products—he is probably the only person in New Zealand who can make that claim—gave House of Travel the competitive edge over other agencies. They essentially used his profile to win the Virgin Galactic appointment. Virgin Galactic conducted a one-day, intensive training in January 2007 for ten House of Travel agents, representing different regions of New Zealand. As Collier recalled, the training covered "everything from the physics involved in space travel, medical, details on the training process that future astronauts would undergo and a lot of aspects on selling."

The agents left the session with a large, comprehensive training manual. ASAS were also invited to do hands-on training, at their own expense, at the National Aerospace Training and Research (NASTAR) Center in the United States, near Philadelphia, something that Collier hasn't yet undertaken because of the large expense of traveling from New Zealand.

Collier came away from the Virgin Galactic training course very enthusiastic. He was pleased that a big global brand was at the forefront of all this. "It's going to happen. It's just a question of when. I can see the momentum growing," he said. "Taking tourists to space is part of the beginning of a new space tourism industry. Rich people in history are always the drivers for any new technology. And the impact that it could have on commercial aviation? I'm thrilled at the idea of getting on the plane in Christchurch and being in London in two hours!"

Two of Virgin Galactic's booked clients were part of the ten New Zealand ASAS. Since then, House of Travel has signed two more, for a total of four suborbital clients so far (as of 2009), giving New Zealand, a country of about four million people, "the highest per capita in the world of suborbital clients," Collier notes. For a nation of adventurers, explorers, and innovators, New Zealanders are just the type of people that suborbital flight would appeal to.

Virgin Galactic started taking reservations in 2005, shortly after the X PRIZE launches. Unlike its predecessors, which were strapped for cash to market suborbital flights, Virgin benefits from its global brand and large, well-funded marketing machine. The company's suborbital booking structure is very similar to Zegrahm's and SA's. Clients wanting to be at the front of the line when flights become available are called Founders and are invited to book the first one hundred seats. Founders pay the full $200,000 upfront and are entitled to numerous perks and privileges, not to mention access to a social network of high-net-worth individuals.

Other categories include Pioneers and Voyagers. Pioneers pay a deposit of $100,000 to $175,000 and are expected to fly within the first year of Virgin Galactic's operations. Voyagers, who pay $20,000 deposits, will follow the Pioneers, estimated to be after one thousand seats have gone up.

Since 2005, public interest in spaceflight has increased dramatically. Virgin Galactic and Richard Branson became familiar names, synonymous with space tourism. Virgin began to get reservations from a mix of professionals serious about space and about preparing for their flights. Those holding coveted tickets for a flight often thought of themselves as pioneers who were helping to launch a whole new industry, a whole new chapter in space travel.

Loretta Hidalgo was a member of Peter Diamandis's team, managing special operations for the X PRIZE launches in Mojave, when she heard about Branson's Virgin Galactic announcement. Thrilled by the prospect of spaceflight since she was a kid, she could not have been more ecstatic.

"After Branson announced it and after having had the first private space-craft in space, it suddenly just became a lot more attainable and real." Little did she know at the time that within a few months, she would have a reservation on a Virgin Galactic flight.

Hidalgo is part of a new generation of young professionals from the post-Apollo era whose enthusiasm and high energy kick-started a resurgence in space advocacy. Growing up in northern California, starry-eyed and ideal-istic, she assumed that by the time she was an adult, everyone would have a rocket in the garage. "My whole life, I just assumed I'll get to go. It was never anything I questioned." She thought about applying to the astronaut corps at one point and had pursued an advanced degree in biology to be as-tronaut eligible, a prerequisite to send in an astronaut application.

"When they [NASA] did the call last year [2007], I promised a friend I would apply. As the deadline grew closer, I imagined going through the panel for the interview. I realized that I wouldn't be able to really say that was what I wanted to do." Training eight hours a day on space shuttle sys-tems was not what she had in mind. "It never dawned on me that I would need to do anything as antiquated as ride a space shuttle," Hidalgo explains. "I wasn't willing to give it up, my advocacy, my outreach, for years, for the opportunity to walk on the moon."

Instead of applying for the astronaut corps, Hidalgo built a career around space science and space advocacy. In 1999 she met her future husband, George Whitesides, at the first United Nations Space Generation Forum in Vienna. Whitesides went to study in Cambridge and returned to work for a start-up company in Pasadena, California, near where Hidalgo was doing her graduate work at Caltech. During her time at Caltech, Hidalgo participated in the International Space University Summer Session pro-gram in Valparaiso, Chile.

Whitesides would eventually become executive director of the National Space Society (NSS) and later was appointed chief of staff to Charlie Bolden, NASA administrator, after serving on the NASA transition team for the incom-ing Obama administration. Together, Whitesides and Hidalgo cofounded Yuri's Night, a worldwide celebration in honor of Yuri Gagarin's historic spaceflight that promoted space awareness to the general public. Today, Yuri's Night is celebrated in more than forty countries each year on 12 April.

While working at NASA Ames Research Center for Mars specialist Chris McKay in 1999, Hidalgo spent two weeks on Devon Island in the Canadian

Arctic, simulating a Mars mission. She spent her time there as a field biologist doing research on arctic plants. In 2000, while at Caltech, she met JPL engineer Michael Eastwood, who would become one of her mentors.

Eastwood brought in Academy Award–winning film producer and director James Cameron to speak at JPL. Cameron was organizing an expedition to the hydrothermal vents in the bottom of the ocean and needed astrobiologists. "I and two of the other JPL astrobiologists were lucky enough to be among the twelve scientists selected to go on the expedition," Hidalgo recalls. The expedition was documented in the 2005 IMAX film *Aliens of the Deep.*

When Diamandis's Zero-G Corporation started its first commercial flights in Florida in 2004, Hidalgo was recruited to serve as flight crew and later flight director. Later that year she moved to Washington DC to join Whitesides and began working at NASA's Office for Exploration Systems.

"We were in London over New Year's, and George had set up a meeting with Virgin, before going to the airport. I was hanging out in the Internet café while George went to the meeting. I was so jealous. Why didn't I set up the meeting with them? I was assuming he would go in to talk to them about some NSS stuff. When he came back, in fact, he had been talking to them about the potential of being the first honeymoon couple for Virgin Galactic." It was definitely not what Hidalgo had been expecting.

On a trip to New York two months later, George proposed to Hidalgo during a visit to Manhattan's Central Park on the occasion of the *Gates* installation, a controversial vinyl and nylon work of art by Christo and Jeanne-Claude. Specific terms for the first honeymoon couple in space were crafted with Virgin Galactic over several months, and the agreement was signed on 12 April 2005. The official announcement of their plans took place at the Experimental Aircraft Association's AirVenture show in Oshkosh, Wisconsin, that year, as they shared the stage with Branson in front of about five thousand people. They both received their Queen's jeweler's Founder's design pins at the event.

"Being a Virgin Galactic Founder has been amazing," Hidalgo notes. "The plan is to have a raffle for Founders (to determine their place in line). Everyone has an equal chance of going first. It's totally random. They're not going to do it for a while. It's actually marketing genius on their part, because it allows them to sell the hundredth ticket with the same excitement as the first. It makes us all first among equals. We're all Founders."

Founders would also serve as glorified guinea pigs. Until Virgin Galactic came along, no one had ever prepared a group of ordinary citizens, an eclectic mix of luxury tourists, for spaceflight. About seventy-seven Founders underwent a two-day centrifuge training exercise at the NASTAR Center, outside Philadelphia. The training included low-g-load runs, followed by high g's on the first day, and then a full SpaceShipTwo flight profile simulation. The centrifuge simulation is complete with the full IMAX-like visual experience.

"Six g's is definitely not comfortable. It's not a ride that I'd recommend everyone to do for fun, but it's an awesome opportunity to get to experience what some of the most grueling spaceflights are like," Hidalgo said. It was not a requirement to pass, but it gave Virgin the opportunity to determine whether regular people off the street, of varying age and medical limits, could endure these forces. Prior to this, there were really no specific medical data available on which the spaceflight medical community could base their flight requirements. Of the Founders who took the training, only a handful would prove to be medically unsuited to fly.

This was Virgin's opportunity to fully test components of their program, to find out what worked and what didn't, using their prime clients as test cases. It made the Founders feel a part of the process of development, helping to shape the program as it moved toward commercial viability.

Being a Virgin Galactic Founder certainly has its privileges. "We got our Puma-sponsored flight suits and Puma shoes," Hidalgo explains. "Definitely makes you feel like a sports hero." Hidalgo and Whitesides have attended five of Virgin Galactic's events, including being flown on a chartered jet to witness WhiteKnightTwo's unveiling in Mojave. They have hobnobbed like movie stars in the Hollywood hills in Bel Air and partied with fellow Founders at New York's Hayden Planetarium for the unveiling of the Space-ShipTwo design. Through these events, they've gotten to know quite a few of the other Founders—all enthusiastic and accomplished, with a "Let's get on with it, let's do it" attitude. At a space event in Croatia, they spent time with fellow Founder, Danish-born Londoner Per Wimmer.

Wimmer has accomplished much during his forty-some years. With blond hair, a charming smile, and dashing good looks, he moves easily in social circles. His Web site, WimmerSpace.com, describes him as "a global financier, an entrepreneur, an adventurer, a pioneer and a philanthropist" — or, as a commenter has characterized him, a "true Indiana Jones meets 007 James Bond."

Wimmer has four master's degrees and speaks several languages fluently. Prior to founding his own London-based investment bank, Wimmer Financial, he worked as an investment adviser for MF Global/Man Securities, Collins Stewart, and Goldman Sachs. In his earlier years he interned at the United Nations and was the only non-U.S. citizen selected for the Presidential Management Intern program under the Clinton administration in 1998.

That same high-adrenaline approach to business also extends to the pursuit of adventure. Wimmer has traveled to more than fifty countries, spending time with the Indians of the Amazon forest, trekking in South America, diving in shark-inhabited waters in Fiji, riding across the United States on a Harley Davidson, skiing at the highest ski resort in Bolivia, and most recently making history for doing the first skydiving tandem jump from above Mount Everest.

Wimmer's interest in space developed in high school but remained on the back burner until one day in 2000. He was talking about possible adventure trips with Lars Plenge, owner of the Danish travel agency Plenge and Co. Plenge was a reseller for SA and had already arranged zero-g flights and space facilities tours for international fashion designer Isabella Kristensen. Plenge had an idea for the travel excursion of a lifetime. For his next great adventure, why didn't Wimmer consider going to space by buying a suborbital flight ticket with SA?

"It took me less than forty-eight hours from the moment I found out to the time I wrote the first check," Wimmer recalls. Considering that SA had not yet flown Dennis Tito to the ISS, and the opportunity for private space flights was not well known beyond the immediate space community, Wimmer was pretty confident of his decision. "I think it was a degree of faith in the beginning. I looked through the materials. It looked pretty real. I thought the idea and concept of going was so exciting that I could not help think that if this turned out to be an option, I'm going!"

To Wimmer, space travel stood for everything he believed in, and the attraction was just too great to resist. When asked about his reasons for embracing the concept of space travel, he said he just liked to be involved in things that are pioneering. "My interest in private space activities and being part of that team that is making it happen is terribly exciting."

He also argues that it's "difficult to be an Indiana Jones on Earth in the 21st century, because the Earth has been discovered. Wherever you go there is always a Coca-Cola dispenser." The true adventurer pioneer nowadays has to look upward, to space. "The job of people like myself who are adventurous is to push the boundary up there to open up space. That is the next stop, the next big thing." When his imagination spins out the full possibilities of this great adventure, he sees himself planting the Danish flag on the moon.

Within nine months of booking his suborbital flight, Wimmer was taking zero-g flights, spinning in the world's largest centrifuge, and flying aboard the MiG-25 jet fighter in Russia. In April 2001 he was one of a group of VIP guests to see Tito launch from Baikonur. "For me personally that was a very big moment. It was a historic moment, and I was there." It would mark the start of his love affair with space, which has spanned close to a decade now. "I was certainly one of the first Europeans to sign up for a private spaceflight opportunity. Because of that, I was able to follow this from the very early days."

During the first few years, it was just a hobby for Wimmer. "It wasn't necessarily something I was talking to people about. In fact only a few people knew about it. They knew that I was a bit of an 'original' and always very different. But first of all they didn't believe it. They thought it was a bit of a joke." That all changed, however, the day that Reuters news agency reported that Wimmer planned to go up in space. It was picked up by sales traders at Goldman Sachs in London, where Wimmer worked at the time.

All of a sudden twenty-two thousand employees received the news in their e-mail inboxes. "Then things really kicked off from there. Attention and visibility led to media interest." Back in Denmark, people began asking Wimmer for interviews and inviting him to give talks in schools. Offers came in from corporations for him to give presentations. "Oh, my God! Well, they wanted to pay for it. I started developing more serious presentation materials. What is really driving me, what can I tell apart from this being a great story about space."

His efforts gave birth to WimmerSpace and the evolution of what he calls his seven fundamental values, the centerpiece of his speeches and motivational talks. His last principle, "Inspire others, especially children, and encourage them to live their dreams," is his driving force for devoting so much time to the project. He notes that he works two full-time jobs, one-and-a-half-time running his investment bank and half-time on WimmerSpace.

Wimmer has been a private spaceflight client long enough to have experienced the sea change in attitude. When he first began his space activities, the press was skeptical. Interviews were tough, because people like Wimmer were considered either dreamers or lunatics. Today, with private orbital flights scheduled twice annually, and with Virgin Galactic's marketing empire promoting Rutan's hardware development, things have changed. There's more a fascination with "the hows and the whys" than speculation over whether spaceflight is possible or not. Currently, Wimmer typically gives two or three media interviews per week. "The interest is definitely there and is not going away."

Virgin Galactic finally caught up with Wimmer in 2007. Virgin was looking for more Founders with certain kinds of profiles, hoping to increase the geographic spread of their clients. "My name is somewhat out there in the space tourism community, because I was one of the first to sign up and that had not escaped the sales people at Virgin." After a number of conversations, Virgin offered him a seat as a Founder.

Having already reserved other suborbital flights didn't prevent Wimmer from booking with Virgin. "I do want to go up on multiple vehicles, and I also want to get up there as soon as possible. It's a way of hedging your bets," he explains. He has now bought three spaceflight reservations, including a flight on the XCOR vehicle. Since then, "I've gotten more than I've bargained for in terms of special events, unique access. It's been a true privilege."

Few prospective suborbital flyers are as prepared for the experience as Wimmer: centrifuge experiences, MiG flights, zero-g flights. He can't remember how many zero-g flights he's done. He says that when most people fly zero g for the first time, "they get excited, they start spinning around, almost like a drunken feeling. They think they can do anything they want. Those are the ones that typically end up throwing up in the background." However, having gone through the training, he knows what it's going to look like at various stages of the flight. "I feel like I'm incredibly well prepared for the actual trip."

Part of that preparation also included a rigorous medical screening that nearly cost him his shot at space. Each Virgin Galactic Founder is required to submit a full medical report on their current health status. Unfortunately for Wimmer, a blip on his ECG concerned Virgin's flight doctor. The doctor wanted more tests performed before he would allow Wimmer to undergo training. Wimmer had to wear a heart monitor for twenty-four hours and get full CT scans. "Turned out everything was absolutely fine. Sometimes people have blips, but it's normal. The good news is that I got a full check and it came out completely clear," Wimmer said, with an enthusiastic smile. In retrospect, he was glad that Virgin took the medical requirements so seriously.

Today, Wimmer is happy to be part of the process for advancing private spaceflight. "I don't like to see myself just as a customer, though I do write checks from time to time. But more than that, I like to see myself as part of a team, helping out whenever I can."

In 2004, after personally witnessing SpaceShipOne's winning X PRIZE flight in Mohave, Wimmer thought he'd be up in space within two to three years. Today he thinks the timeline for flight is a "moving target." Virgin emphasizes the importance of safety above all else. So Wimmer is content to wait. He eagerly anticipates each milestone debut and scheduled flight test that Rutan and his Scaled Composites crew put out from time to time. Having been in the system almost a decade, he's in for the long haul and having fun in the process.

As of May 2008, Virgin had eighty-four Founders and had closed the list. However, clients continued to sign up as Pioneers and Voyagers, who would take later flights. By the following year, three hundred paying customers from thirty-nine countries had paid $40 million in flight seat reservations. Proof that interest in space travel has penetrated the mainstream were bookings from such high-profile celebrities as actress Victoria Principal, *Spiderman* director Brian Singer, and designer Philippe Starck.

Starck was commissioned by Branson to put his unique stamp on Virgin's space business. First he created the Virgin Galactic logo, which prominently features Branson's own eyeball. Then he designed part of SpaceShipTwo's interior and the spaceport that will be the center of Spaceport America, the launch facility being built in New Mexico.

As Virgin realized the importance of the Founders, it became more se-
lective in whom it allowed onto the list for the first flights. Individuals were
invited to become Founders only after being carefully selected for their
geographic spread and occupational diversity. They come from all over the
world: Ibrahim Sharaf, president of the Sharaf Group in the United Arab
Emirates; Trevor Beattie, an advertising guru from the United Kingdom;
Namira Salim, an artist originally from Pakistan; Princess Beatrice, fifth
in line to the British throne; Sunil Paul, a San Francisco green technology
entrepreneur; and Mark Rocket, an Internet entrepreneur planning to offer the first launch services in New Zealand.

For those who were not fortunate enough to get one of the coveted spots
on the Founders list, not all is lost. Spots for Pioneers and Voyagers are still
up for grabs for a minimum deposit of $20,000. John Criswick did just that
and became the first Canadian to book a flight with Virgin Galactic.

Criswick was not new to the space industry. He spent most of his early
years dreaming of becoming a Canadian astronaut. After earning degrees
in electrical engineering and space physics, he worked as a software engineer for Canadian Astronautics in Ottawa and at an observatory in Utah.
He kept his space dreams alive by attending the International Space University Summer Session program in Toulouse, France, in 1991. He applied
for the Canadian Space Agency's astronaut program, even though he knew
it was a long shot. "There were 5,200 applicants for six positions. I didn't
have enough PhDs," Criswick jokes. His path toward spaceflight did not
materialize then, but as luck would have it, his foray into the tech business
world gave him the means to put down a deposit for his space flight.

When a company Criswick founded was purchased by Sun Microsys-
tems, he earned $20 million. He's still running a dozen or so tech compa-
nies in Ottawa, but after he'd made his millions, he decided it was time to
return to making his dreams of space flight a reality.

In January 2008 Virgin Galactic unveiled the model designs of SpaceShipTwo
and WhiteKnightTwo at a New York event to an eager crowd of media reporters and Virgin Galactic ticket holders such as Hidalgo, Wimmer, and
Criswick. Whitesides, who by then had been named Virgin Galactic senior adviser, was on hand as host for the event. Virgin Galactic's team was
represented by Branson and Galactic CEO Steve Attenborough. Rightfully

sharing the spotlight this time, along with Burt Rutan, were Jim Tighe and Bob Morgan, chief project engineers for SpaceShipTwo and WhiteKnightTwo. They were given the honor of explaining the features of the new spaceflight system that was to take commercial passengers into space.

The new spaceship and mother ship were considerably larger than their predecessors but had the same distinctive elegant design. WhiteKnightTwo will be the largest 100 percent carbon composite airplane in the world, with a 140-foot wing span, equivalent to that of a B-29 bomber, and a coast-to-coast ferry range. It will use four rather than two Pratt and Whitney PW308 jet engines. Its unusual design includes two identical cockpit cabins — almost like two aircraft joined by the wing.

The cabins will serve as a training facility for passengers and pilots, as well as a potential research facility for scientific payloads. It can serve as a flying centrifuge for zero-g- and high-g-load training on parabolic flights. Family members of spaceflight passengers could conceivably ride on WhiteKnightTwo to view their loved ones on SpaceShipTwo rocket into space.

SpaceShipTwo is also significantly larger than SpaceShipOne. It can now carry two pilots and six passengers up to a height of sixty-two miles. Tighe explained that he is six foot three inches tall but can easily walk upright with additional room between his head and the ceiling. The craft's eighteen-inch panoramic windows will allow passengers to view the curvature of the Earth below, a span of about one thousand miles, from the Gulf of California to San Francisco Bay, as they float in weightlessness for close to four minutes at maximum altitude.

The flight will last about two hours from take-off to landing. Released at about forty-eight thousand feet by the WhiteKnightTwo carrier ship, SpaceShipTwo will use twin hybrid rocket motors to climb at three times the speed of sound to peak altitude.

Seats will fully recline to maximize the cabin space during zero g and allow passengers to float around the cabin for a full flight and visual experience. Passengers are expected to wear body-hugging futuristic pressure suits as a precaution within the "shirt sleeve" environment. They will experience three and a half times the pull of gravity on the way up and a few seconds of 6 g on the way down. At eighty thousand feet, SpaceShipTwo will begin its glide and return to land at Mojave Airport, utilizing the same unique feathering mechanism as SpaceShipOne.

The spaceflight will be the dramatic conclusion to a three-day training program, details of which are still being finalized. Virgin is developing the training using the experience and simulator data acquired through the training sessions at NASTAR. Out of eighty Founders tested, only two were not able to take the centrifuge ride because of medical disqualification and had to withdraw their reservations. Three more were asked to delay training for treatable conditions.

This represents a 93 percent success rate, a much higher number than Virgin had anticipated. Such findings undercut the long-held belief that only professional astronauts could travel in space. James Lovelock, founder of the ecological hypothesis known as the Gaia theory, is eighty-eight, and Branson's son James is age eighteen. To date they have been the oldest and youngest to undergo training. According to Attenborough, this proves that even "ordinary people can actually go in space and that almost all of us have the right stuff."

Rutan expects to fly passengers to space twice a day, while WhiteKnight-Two can fly three to four times a day, allowing it to be used for other scientific or training purposes. It is expected that 80 percent of Virgin's market will initially come from space tourism during the first seven to eight years, with the goal of flying five hundred people in the first year of operations, and a total of fifty thousand in the first ten years.

But Virgin's business plan doesn't end in space tourism. White Knight's high-altitude lift capability opens up a host of other space applications, including serving as a high-altitude small satellite payload launching facility and perhaps as a launching pad for human-rated orbital flights in the future. "We don't have the funds to develop unmanned satellite launchers," Whitehorn notes, but he feels it is only a matter of time before some other vehicle developers begin to look at the potential market.

Rutan expects to build forty to forty-five SpaceShipTwos in the first twelve years, capable of flying about one hundred thousand people over that time period. Virgin is already looking at suborbital science research opportunities that could drastically cut the cost for Earth science, astrophysics, or biomedical research through flexible and customer-driven timing and location requirements. The fact that experiments can be flown together with manned supervision could greatly cut the preparation time and design cost for payloads. Space qualification of instruments and equipment is another market that could easily be served by SpaceShipTwo flights.

For the unveiling of WhiteKnightTwo on 28 July 2008, Virgin Galactic flew a group of Founders from Los Angeles to Mojave Airport on a new Virgin Airlines Airbus. The nose of the Airbus was painted with the message, "My other Ride is a Spaceship." The Founders witnessed the fanfare as Rutan and Branson, wearing matching white shirts with prominent Virgin Galactic logos, waved to the crowd from WhiteKnightTwo's cockpit windows. They taxied the plane in front of Scaled Composites's hangar, where the VIP guests had gathered.

Named *vms Eve* by Branson, WhiteKnightTwo was dedicated to his pioneering mother, Everett Branson. The craft began test flights on 22 December 2008 and continued them throughout 2009, increasing its range and altitude capability each time. It debuted before the general public at the Oshkosh air show in July of that year.

SpaceShipTwo's first rocket-powered flight is planned to occur sometime in 2010. It will then go through further extensive testing until Virgin determines that the spacecraft is ready for commercial flight. "Safety is our biggest priority in this project," Whitehorn emphasizes. It is a claim borne out by the ambitious planned testing program. With over two hundred flights, it will be the largest test flight project ever undertaken, bigger even than that for the Concorde jetliner.

Branson is abundantly confident in the ultimate safety of the flights he will offer to space. In fact, along with Burt Rutan, Branson announced that he has signed on for the first flight. "My parents, and my two children Holly and Sam, and I, will be the first passengers on the inaugural flight of SpaceShipTwo."

The power of such gestures to spur this fledgling industry was recognized in 1997, in the Space Act Agreement Study, underwritten by NASA. It made the bold suggestion that one way to hasten the development of the potentially huge market in space-related business would be for our "space leaders" to take trips to space.

In acting on that advice, Branson's own flight to space may go a long way to assuring an emerging industry that the time is right to embrace the technology and the opportunity to deliver on the promise of private manned spaceflight.

10. It Takes More Than a Spaceship to Build an Industry

I think it's going to be a wild ride the next
twenty years as this industry emerges.

Stuart Witt, general manager,
Mojave Air and Spaceport

Nearly forty years ago, Gerard O'Neill changed attitudes toward the private space movement by giving it academic credibility and encouraging the development of new technologies and new policies that would support human communities in space. The space shuttle, electromagnetic accelerator, solar power satellite, and the Moon Treaty were but a few of the big pieces that had to fall into place for space colonies to become a reality. In the end, the failure to realize some of these basic elements doomed further progress on the colony movement.

The challenges to develop private manned spaceflight are far less daunting than those of creating orbital cities, but success still depends on key developments. If someone were to make a list of necessary requirements that have already been accomplished, it would include the following: (1) private investors visionary enough to finance innovation, (2) a suborbital space vehicle to jump-start space tourism, (3) tourism companies marketing space travel opportunities, and (4) a cadre of wealthy individuals willing to pioneer space travel.

The items that still need to be added to that list: (5) launch facilities, (6) government regulations that encourage the industry, (7) a private heavy launch vehicle to ferry freight and passengers to space, and (8) a space destination. How soon private individuals travel to space and how many of them have that opportunity depends on meeting these objectives. The future of private space flight is being written right now, in a hundred different venues, with a hundred different technologies, by a hundred different individuals.

Although overshadowed by Mike Melvill's historic SpaceShipOne flight on 21 June 2004, an equally significant event occurred in Mojave just four days earlier: the Mojave Airport obtained its launch site operator's license from the FAA. That official stamp of approval transformed the airport into a spaceport, the world's first civilian airport-spaceport combination. The license authorized the facility to conduct horizontal launches of reusable launch vehicles (RLVs) as well as perform rocket engine testing and vehicle manufacturing. This authorization covered air-launched vehicles such as SpaceShipOne, as well as spacecraft capable of launching from a standard runway like XCOR Aerospace's planned Lynx vehicle.

Receiving this designation was a giant leap for dusty Mojave Airport, which had begun its existence in 1935 to support the local mining industry. It later hosted the Marine Corps gunnery school for World War II pilots and still serves as home for the National Test Pilot School. But nowadays it is considered the Silicon Valley of the aerospace world, home to the headquarters of forty-some companies engaged in highly advanced aerospace research, design, and development.

Prior to launch licensing, Mojave Airport already had as tenants such RLV companies as Rotary Rocket, Scaled Composites, XCOR Aerospace, and Inter Orbital Systems, and it continues to attract the best and brightest in the commercial space business to set up shop in its facilities.

Because Mojave was the first civilian spaceport considered for licensing by the FAA, it suffered through the long and tedious process of working with the government to define exactly what a civilian spaceport was and how it would function. Under the leadership of Stuart Witt, Mojave Air and Spaceport's general manager, Mojave paved the way for other nonfederal spaceports with a focus on commercial RLV flights.

It took Witt, a former Navy and test pilot, about five years to obtain Mojave's spaceport operator's license. In cooperation with the FAA's Office of Commercial Space Transportation, under Patti Grace Smith, the groundbreaking process involved environmental assessment and public safety issues. The process included revisiting statutes regulating propellants, defining spaceport operations boundaries, and understanding the complex problem of defining a route, airspace, and flight plan for suborbital launches.

Witt admits to having become a little frustrated with the bureaucracy at times, especially when it came to certain nonaerospace issues, such as watching out for *Gopherus agassizii*, the endangered local desert tortoise. "At 300 takeoffs and landings a day, nobody asked us to ever do a tortoise check. But before I can clear a spaceship to land, I have to do a tortoise check of the primary runway."

Tortoises aside, the licensing procedure created a learning process for both the Mojave and the FAA office teams, just as it had for licensing SpaceShipOne, as they addressed all of the new issues concerning commercial spaceports.

Now that Scaled Composites has rolled out both WhiteKnightTwo and SpaceShipTwo, it is anticipated that flight testing and rocket development will continue at Mojave through SpaceShipTwo's maiden flight, perhaps a two-year process. A few noncommercial passenger flights then follow, also at Mojave. However, the honor of actually conducting the highly anticipated commercial flights for Virgin's Founders belongs to another spaceport on the rise, deep in the New Mexico desert.

On 31 December 2008 Virgin Galactic signed a twenty-year lease with the state of New Mexico, planting its base of operation and global headquarters in Spaceport America, the first purpose-built commercial spaceport in the United States. Located in Las Cruces near White Sands Missile Range and Kirkland Air Force Base, Spaceport America had received its launch license from the FAA just two weeks earlier. This gave the green light for construction of the spaceport's planned facilities, including the futuristic terminal and hangar facilities, created from the fertile imagination of world-renowned designer Philippe Starck.

Once the environmental issues were cleared and plans approved by the state, ground breaking for the multimillion-dollar, state-of-the-art Virgin facility began in June 2009. "We are trying to synch our construction cycle with the Virgin development cycle of their vehicle [SpaceShipTwo]," explained Steve Landeen, Spaceport America's executive director. They expected to finish and be operational by 2010.

"Is this the next Wright Brothers?" Landeen can't hide his enthusiasm. "I believe it potentially is. Tourism folks are really technology investors, just like the early days of aviation."

Landeen took over the ambitious project in January 2008. Having worked for Honeywell and Landmark Aviation for twenty-three years, he banks on his years of combined business and technology experience in strategic planning, project engineering, and business development to lead the project to its fruition.

Already other private companies, such as Up Aerospace, Lockheed Martin, Armadillo Aerospace, Microgravity Enterprises, and the Rocket Racing League, have signed agreements to take up residence at the Spaceport. This would require separate buildup schedules for vertical launch and manufacturing facilities from Virgin Galactic's terminal.

Although initially dedicated to a purely space tourism business model, Spaceport America plans to expand its operations to partner with the U.S. Air Force for military applications involving surveillance, reconnaissance, and balloon development, as well as satellite launches on human-powered flights. It is already talking with NASA about potential cooperation on scientific flight applications. For scientific purposes, Virgin flights would be a good deal. Less expensive than a sounding rocket, they would also allow for faster payload development and for scientists to accompany their experiments into space.

An educational institute and a space theme park are also in the works. Point-to-point suborbital transportation is already under consideration for cargo transport and operations response activities for the Air Force and the Marines. The "age of the classic, high-consuming power vehicle is going to get transformed into this new agile vehicle that can deploy systems," Landeen said. He predicts that all these applications could eventually eclipse the space tourism business, depending on how the market develops.

A network of spaceports, serving earthly travelers, is the next logical step in spaceport development. Just as the early airports in aviation developed as strategic nodes in a global transportation system, interrelated and functioning in collaboration, not in competition with each other, spaceports will eventually offer travelers point-to-point transport at the speed of space flight.

Unlike space tourism flights, which will launch from and land at the same location, point-to-point, suborbital flights will launch from point A and land at point B. They could, for instance, reduce to two hours the flight time from Washington DC to Tokyo. If this high-speed, global transportation of passengers and cargo could grab even a small piece of the commercial aviation market, it would be far more lucrative than sending tourists into space.

With that purpose in mind, Spaceport America has signed a partnership with Spaceport Sweden, another spaceport hub, for future Virgin Galactic flights. The sister spaceport relationship is a natural collaboration for establishing transcontinental suborbital landing ports for Virgin's future global spaceport infrastructure. A spaceport in Scotland is being proposed as another potential launching site for Virgin's future global fleet.

Just as each town and city invests in transportation as a natural barometer for its success and growth, a number of states have boosted their viability to host spaceports and are seeking government support and incentives to entice commercial business. Federal launch facilities at Kennedy Space Center and the Cape Canaveral Air Force Station, as well as Wallops Flight Facility (near Chincoteague, Virginia) and Vandenberg Air Force Base (Santa Barbara County, California), have each developed their nonfederal counterparts to attract commercial tenants.

Spaceport Florida signed agreements with SpaceX, Bigelow Aerospace, and PlanetSpace and is in the process of licensing more launchpads for commercial operations. On the other hand, individual state facilities, such as the Oklahoma Spaceport and the Kodiak Launch Complex in Alaska, are doing their own promotions. Oklahoma Spaceport acquired its launch operations license in 2007 with RLV companies Rocketplane and TGV Rockets as anchor tenants (Rocketplane pulled out of Oklahoma Spaceport in February 2009).

States use local government tax policy incentives to attract aerospace and suborbital entrepreneurial companies to make maximum use of previously existing investments already in place. The state of Hawaii is looking into creating multiple spaceport facilities at Kona and Oahu airports, potentially the first FAA point-to-point application. Alabama, Texas, Wisconsin, and Wyoming are in different proposal stages, some with pending license applications with the FAA.

On the international front, proposals and concepts for spaceports have been put forward in Singapore; Hokkaidō, Japan; Montpellier, France; Barcelona, Spain; Dubai; and the United Arab Emirates. Concepts for spaceports on the islands of the Bahamas and Curacao take advantage of equatorial proximity in trying to convince their local governments of the viability of a lucrative space launch business. Others, such as the Woomera launch site in Australia, have existing facilities in place waiting for commercial developers from abroad to set up shop.

For the present, each of these spaceports counts on regional and local appeal for the paying passenger who would like his or her spaceflight viewing experience to be somewhat familiar. According to Rocketplane's Chuck Lauer, there will be a "Green Planet" and a "Blue Planet" view of the world. "What they see from any given vantage point is the value proposition."

All these spaceport developments paint a bright future for commercial spaceflight, but progress won't happen unless there is sufficient funding, the right mindset to succeed, and governmental backing that can put new rules and regulations firmly in place. The United States, with entrepreneurial companies slowly emerging and a whole infrastructure being created, maintains its global lead by trying to create policies that keep pace with this rapidly developing industry.

The Space Flight Act Amendment, signed in 2004, empowered the Office of Commercial Space Transportation (AST) to adapt certain requirements and liability policies that would enable launch operators to fly participants by informed consent and on an "at your own risk" basis. This allows U.S. launch developers to get into the business without spending billions of dollars in vehicle certification. The FAA has also mandated nonrestrictive medical, flight crew, and passenger requirements, allowing operators to hasten development of vehicle life support systems and passenger training to meet federal guidelines.

AST has taken on a supportive and encouraging approach while maintaining its safety mandate, and is breaking new ground in terms of commercial launch and operations licensing. "Don't overregulate this industry," AST administrator Patti Grace Smith has cautioned. "If we regulate the industry the way certification would require—all the vehicles to be certified, with all the tests and costs—the industry will never get off the ground."

You don't have to look too many years into the past to realize what a sea change in attitude this represents for the government. Both the FAA and NASA have redefined their way of doing business to enable the development of the private space industry. NASA, not so supportive at the time of Dennis Tito's launch, has since changed its attitude about commercial space, starting with approval of the ISS spaceflight participant requirements that allowed private participants access to the ISS.

Today, NASA uses commercial companies for its own needs. Zero-G Corporation, owned by Space Adventures, now flies all NASA parabolic flights out of Johnson Space Center. NASA has awarded SpaceX and Orbital Sciences Corporation contracts for the development of ISS cargo resupply and is looking into commercial RLV development applications for scientific purposes. It is anticipated that NASA will contract out development of a new commercial crew transportation system for the ISS to commercial companies such as SpaceX.

Globally, other nations are beginning to take an interest in the private spaceflight business. In the United Kingdom a few small private companies emerged in the wake of the SpaceShipOne launches. Others, like Star Chaser, Bristol Aerospace, and Reactions Engines Limited, predate the X PRIZE competition but have long been stuck in the design and prototyping phase because of lack of funding.

In recent years, the European Space Agency and the European Union have commissioned studies on commercial human spaceflight to gain an understanding of all the technical, operations, and legal issues needed to start a space tourism industry. The large aerospace corporation European Aeronautic Defense and Space Company (EADS) has put forward proposals for suborbital vehicle concepts but has yet to commence actual hardware development.

Small companies such as the Austria-based Orbspace, founded by aerospace engineer Aron Lentsch, have learned to adapt to the European conservative culture on space tourism. Orbspace is studying rocket propulsion technologies and suborbital operation and safety aspects, funded by the Austrian Space Agency and the European Union. According to Lentsch, these small development and research projects enable Orbspace to build its capabilities with public funding. This will position it to eventually work on an experimental vehicle.

Having worked in propulsion systems and future launch vehicle studies for the European, Japanese, and French space agencies, Lentsch believes the only way that small companies can succeed in Europe is by designing simple, uncomplicated vehicle systems that have smaller development costs and proven reliable systems. Orbspace hopes to develop its Infinity vehicle, a small vertical takeoff and landing capsule designed to reach two hundred kilometers carrying three passengers.

Other European companies making similar efforts one small step at a time are the Aerospace Institute and Talis Institute in Germany, the Romanian rocket group ARCA, and the Swiss Propulsion Laboratory, based in Langenthal, each of which has suborbital and orbital manned development aspirations.

But are all of the pieces in place for the European market to match the pace of private space development in the United States? Lentsch thinks not. The scarcity of venture capitalists in Europe paints a bleak prospect for the next step—finding funding for actual vehicle development in the near future. "In Europe, the thinking is only big companies can do it," Lentsch claims. "Nobody believes that small companies can produce rocket engines." Therefore, nobody will invest in a high-risk project such as a suborbital vehicle.

It may well be that at least for now, the market numbers just don't add up. For example, EADS estimates that it would require $2 billion to develop its spaceplane project. If that were the case, recouping the development cost would require a number of flights that exceeded the market demand forecasted by the Space Tourism Market Study conducted by the Futron Corporation in 2002.

Compounding those problems, Europe has not developed the legislative and regulatory framework to support private space travel. In fact, Lentsch believes that given Europe's tendency to overregulate, "it might be too conservative to design a legal framework to give us the chance to try to see if these vehicles are going to work." The European Space Aviation Authority has initially indicated that certification is the best way to ensure regulation and safety for any suborbital vehicle to assume commercial flight, a much more restrictive approach than its U.S. counterpart's in the FAA.

If this happens, "then there would be no other choice but to operate this business outside of Europe," Lentsch concludes. It would affect Virgin

Galactic flight plans at Spaceport Sweden or Scotland, or any other space-port in Europe. Lentsch is hopeful that this attitude will change once a more thorough study of the industry is conducted.

A different scenario is playing out in Japan, where interest in private space activities far outpaces private space development. Like Europe, Japan has been very conservative when it comes to manned space activity and space tourism. Although various design concepts and research and market stud-ies were conducted on space tourism in Japan before the idea was taken se-riously in the United States, Japanese aerospace companies have only just begun to work on developing suborbital space vehicles.

"It has been very delicate to talk about manned space vehicles, including space tourism," explains Misuzu Onuki, a well-known aerospace business consultant, author, and space tourism advocate in Japan. However, that may change under Japan's new Space Basic Plan, issued in 2009.

The Space Basic Plan outlined a strategy for Japan's utilization of space and the fostering of space industries. Announcements followed quickly from the Japanese Aerospace Exploration Agency about plans to recruit ad-ditional astronauts for the first time in ten years and to develop such bold projects as a solar power station.

Onuki believes that Japan's new emphasis on space will herald a change in attitudes about private space development. Even though her role as an independent aerospace business consultant is still highly unusual, Onuki now provides consulting services to the major Japanese aerospace compa-nies as well as to venture projects. "Recently, the mainstream aerospace community has shown an interest in the nontourist applications for sub-orbital vehicles, such as microgravity research and small satellite launch." She thinks that this may motivate the Japanese government to finally pro-vide registration and guidelines for private suborbital craft for space tour-ism and other applications, just as the FAA/AST did in the United States. However, she is not holding her breath, pointing out that "Japan is very slow with regulations."

Despite the lack of progress in private vehicle development, Japan's unique strength is its end-user potential for space tourism. There is a great deal of consumer enthusiasm in Japan for flight opportunities in space, perhaps more than in any other country. Japan's mainstream travel agencies, such as Spacetopia, Club Tourism, and JTB, pioneered selling ISS space tourism, suborbital, zero-g flights, and even the lunar trip.

Onuki is exploiting that enthusiasm by offering such niche space opportunities as Space Wedding. For $2.4 million, the program would allow couples to exchange their vows at one hundred kilometers above the Earth. She also organized Space Couture Design Contest, a nationwide competition to design original fashion that will become part of Rocketplane's official spacewear.

To support the eventual development of space tourism, several regional airports, including those on Japan's northern island of Hokkaidō, have proposed spaceport concepts in an attempt to lure foreign vehicle developers to use their existing flight facilities. However, Japan has not yet begun to sort through the regulatory process that would allow this development.

Elsewhere in Asia, space very much remains a government-run initiative. Russia continues to monopolize private orbital spaceflight opportunities, despite increased demand for NASA crew transport to the ISS once the shuttle retires in 2010. In 2007 and 2008, the Russians launched Malaysian and Korean spaceflight participants to the ISS on Soyuz rockets.

Government space programs in both China and India are aggressively pursuing manned space initiatives. Chinese taikonauts (officially known in Chinese as *hang tian yuan*) conducted successful space walks in 2008, while the Indian Space Research Organization has been given the green light and budget approval to undertake a manned mission to the moon.

However, space tourism has not altogether captured the public imagination in these countries. In a 2007 Beijing presentation, Zhang Qingwei, president of the China Aerospace Science and Technology Corporation, hinted that the concept of space tourism is interesting to the Chinese as a future direction but is beyond the realistic bounds of their cultural and mental mindset in the near term, and also beyond their current technology.

There is probably no one more self-assured and confident of developing the necessary technology for further progress in the risky business of private space than Elon Musk. And why wouldn't he be? His every entrepreneurial venture has turned to gold. He is the type of highly motivated individual who has driven so many of the critical developments in the private space industry.

Since emigrating from his native South Africa to North America when he was seventeen, Musk has been on the fast track to success. After earning degrees in physics and business, he cofounded a company named Zip2 that he eventually sold to Compaq Computer for $307 million. He repeated that scenario when he cofounded PayPal, the largest online banking service. Musk was the largest shareholder when it was sold to eBay for $1.5 billion in 2002.

Having pushed the bounds of the Internet, he moved on to electric vehicles. As current chairman, CEO, and product architect of Tesla Motors, his goal is to mass-produce the first fully electric sports car, the Tesla Roadster. He also serves as chairman of Solar City, one of the largest installers of solar panels.

But currently his primary job is running Space Exploration Technologies, or SpaceX, as founder, CEO, and chief technology officer. As bold as any of Musk's past entrepreneurial endeavors, SpaceX is attempting to simultaneously improve the reliability and reduce the cost of space access — both by a factor of ten. Musk has already invested more than $100 million of his own money to seed SpaceX's development.

Musk doesn't shy away from big ideas. He conveys the impression that he cannot be fazed by anything. For a man who still possesses unusually boyish looks, he sometimes comes across as the quintessential techno-nerd. He seems to relish the challenge of working out propulsion and complex space problems in front of you, spitting out numbers and facts without hesitation and without a care for the time or the tediousness of the exercise.

In an interview with one of the authors of this book at the 2008 National Space Society conference, he explained how SpaceX's continued milestone goals and development could one day lead to interplanetary travel. He used a von Braun quote — "Hydrogen is the reason you can go to the moon" — to explain the importance of achieving a hydrogen propellant upper stage as the key element for success. "With that development you can do a manned mission to Mars," he suggests, which is just what he is trying to achieve. Musk acknowledges that he doesn't have an actual timeline but adds, "Could it be done within the next twenty years? Yes, I say highly likely within the next twenty years." Then he adds, with a smile, "Provided SpaceX is still around."

Unlike other key figures in the development of private space, Musk had no youthful infatuation with the cosmos. "This is not a result of a childhood epiphany. Why would I have a childhood epiphany? Because I watched *Star Trek*? That's kind of silly. My interest in space stems from thinking about what are the important problems facing humanity and life itself. The extension of life to multiple planets seems to be actually the most important thing that we could possibly do." The next big step for mankind is to be a multiplanetary species, and Musk thinks he's got the talent and the money to make it happen in our lifetime.

Unlike most visionaries, Musk is more sobered by looking back at the challenges already met, than looking forward at those to come. During the interview, he described the difficult process of rocket development. "It's a huge complicated system with a passing grade of 100 percent. It *is* rocket science." He confessed that "six years ago, I didn't know anything about rockets." Back then SpaceX's strategy was to make small rockets, learn from them, and make bigger versions of their prototypes. He admits that was something less than a brilliant strategy. "The first two years of its [SpaceX's] existence should be discounted on the basis of idiocy."

Along with the idiocy came the challenges. For instance, in order to significantly reduce the cost of his launch vehicle, Musk had to build 80 percent of his rocket from scratch — "soup to nuts," as he puts it. Red tape for obtaining a launch license forced him to relocate his launch operations to the Kwajalein Atoll in the Pacific. His first three unsuccessful launch attempts were beset with technical problems that prevented the rocket from achieving Earth orbit.

Finally, on 28 September 2008, SpaceX's Falcon 1 successfully launched to low Earth orbit on its fourth attempt. The launch vehicle is capable of lifting 926 pounds of payload to orbit and is partially reusable. SpaceX has signed on nearly $2 billion worth of payload launch business and has been cash positive since 2007. Falcon 1 successfully launched its first commercial payload, Malaysia's RazakSAT, into orbit on 13 July 2009.

Setbacks aside, SpaceX has achieved milestones only a few governments and big aerospace companies have previously achieved, and has done so in less time than one would expect from a start-up company with a modest budget. With the new NASA directive to retire and replace the aging shuttle fleet by 2010, SpaceX is under contract with the agency to provide cargo transport to the ISS. SpaceX plans to have its heavier lift Falcon 9 ready to launch by the end of 2010. Capable of lifting twenty-three thousand pounds of payload to low Earth orbit for about $49.5 million a launch, it will usher in bigger and better possibilities for cheaper launch access to space.

Falcon 9 will also carry SpaceX's Dragon, a reusable spacecraft capable of transporting cargo and crew to the ISS. The first demonstration mission for both Dragon and Falcon 9 is expected in 2010. For Musk, the success of Falcon 9 will bring him closer to his ultimate goal of having the launch capability to send humans to Mars and beyond.

Musk is pretty confident that in the long term, the price for launching to space, be it humans or cargo, will eventually go down because of efficiency and reusability, perhaps to as low as $300 per pound. He is attempting to make the Falcon 9 the first fully reusable launch vehicle. "If you cannot make reusability work, then forget it. Reusability is paramount in any mode of transportation."

Confronting the tricky issue of reusability that has confounded generations of space engineers, Musk's mind kicks into gear. He proceeds to explain that the lowest price point one can expect from a human commercial orbital flight can be derived by comparing the operations and propellant cost for his seven-person spacecraft with a four-hundred-person-capacity Boeing 747 airplane. Cost is a factor of the propellant-passenger ratio, which is sixty times greater for a spacecraft than for a commercial jet.

Insofar as each stage of the three-stage spacecraft corresponds to a separate jet, it would be equivalent to the cost of operating and maintaining three airplanes. Therefore, "the absolute lowest price per person would be a few hundred thousand dollars, and it would be heavily dependent on the flight rate," he explains matter-of-factly. "In order to get there, it must be affordable. If it's not affordable, it's not possible."

Musk is also interested in sending people to low Earth orbit and perhaps participating in the tourism market, but he remains cautious over the notion of spaceflight as a "joy ride." Although he was one of the first to buy a suborbital ticket from Virgin Galactic and thinks "it's cool," he doesn't foresee suborbital flight as a means to an end. "It's not a stepping stone," he confidently notes, adding that the "architecture of Scaled Composites can't evolve to orbital capabilities."

While the giant aerospace companies develop their own launch vehicles, as do EADS and some Russian ventures, Musk has taken the lead among the new commercial companies in developing orbital capabilities. For now at least, much of the future of space commerce and tourism seems to rest on Musk's shoulders.

The success of Falcon 9, and ultimately the Falcon 9 Heavy, would open up other opportunities, such as robotic probes and landers on the moon's surface. Musk has pledged discounted launch opportunities for registered competitors of the Google Lunar X PRIZE, a competition to land the first private spacecraft on the moon. He is also offering an alternative commercial configuration of his Dragon spacecraft called Dragonlab, capable of pressurized and unpressurized cargo delivery and recovery, as well as of serving as a space platform for private and scientific uses in low Earth orbit.

If Musk gets around to launching a space platform, it will not be alone in space for long. An entire industry of would-be space real estate entrepreneurs has been waiting on the sidelines for a reliable and affordable commercial transport system that would make privately owned space stations the new luxury destination of the future. But even before a commercial heavy launch vehicle becomes available, innovative developments in space platforms are heralding a space real estate boom. Private space platforms are now poised to grab a slice of the business that until now has been monopolized by the Russians.

"That's one small step for Bigelow . . . one giant leap for entrepreneurial space," announced Bigelow Aerospace's Mike Gold. The lofty pronouncement came on 12 July 2006, when Bigelow Aerospace, a Las Vegas company owned by Budget Suites mogul Robert "Bob" Bigelow, launched its Genesis I space habitat into orbit aboard a Russian Dnepr rocket. The rocket launched from ISC Kosmotras Space and Missile Complex, an active military base next to Yasny, Russia, near the border with Kazakhstan.

Genesis I was the first expandable space habitat technology ever launched into orbit. A subscale model, 11.5 cubic meters in volume, Genesis I is the precursor to Bigelow's planned full-size habitats that would someday form a crew-rated, privately owned space station in orbit.

Establishing a budget space station for wayfaring astronauts made perfect sense to a man who spent three decades in the real estate and construction business before founding Budget Suites of America in 1999 to serve a niche market between the hotel and traditional rental apartment business.

Bigelow will sometimes trace his space roots back to his youth in Las Vegas, Nevada. It was the 1950s, and a boy's imagination could be stirred by the aboveground nuclear test explosions at the nearby Nevada Test Site and the subsequent alarming light shows they created. His grandparents populated his fantasies with aliens when they spoke about their personal sighting of a fast-moving, brightly lit red object in the sky above the Nevada desert. However, the unexpected career leap to private space entrepreneur came after he read an article in *Air and Space* magazine describing a new inflatable technology being developed by NASA for the ISS.

The idea of inflatables has been around for a long time. In the 1960s, NASA's Langley Research Center had looked into the technology and had a module made by Goodyear. It resembled a giant tire. The design had serious problems and was shelved until the development of stronger, more flexible materials such as Kevlar came along. In the early 1990s the concept was revived to support the Space Exploration Initiative announced by President George H. W. Bush and developed for crew transfer habitat purposes for Mars missions. The inflatable habitat was then called Transit Habitat, or TransHab.

But not all good ideas find their way into NASA's budget. TransHab was put on the shelf until a few years later, when NASA engineers repackaged it to become the planned crew habitat for the ISS. Development progressed until it was eventually canceled because of delays and mounting costs. It was then that Bigelow moved in to exploit the technology for commercial purposes. He founded Bigelow Aerospace in 1999.

Unlike such companies as Barcelona-based Galactic Suites that plan to use inflatables specifically for orbital space tourism, Bigelow envisions multiple commercial uses, including microgravity research, particularly for pharmaceuticals and biotechnology; and the development of an international astronaut corps. There is certainly the potential for serving the space tourism market, but not as the "space hotels" that have sometimes been portrayed in the media.

The launch of Bigelow's Genesis I was also the first time a single integrated payload was launched aboard the Russian Dnepr rocket and the first commercial launch ever to take place at the Kosmotras Space Complex. Other than Bigelow himself, no one was happier about the success than his launch campaign manager, space lawyer and head of Bigelow's Washington DC office Mike Gold. Gold orchestrated the Genesis I project from birth to launch and was at Kosmotras mission control, relaying the countdown process and results via Skype to Bigelow Aerospace's head office in Las Vegas. For their part, the Russians did not disappoint, delivering the space habitat to within four hundred meters of its planned orbital parking space.

It took Gold, under Bigelow's leadership, about half a decade to pull off launching the first privately owned module into space. The effort was made possible by a unique technology transfer agreement with NASA and was nearly suffocated by the United States' antiquated International Traffic in Arms Regulations (ITAR). Bigelow partnered with Russians and Ukrainians to refit cold war nuclear ICBMs as low-cost launch vehicles. In some respects this international collaboration was reminiscent of previous private space agreements hammered out by MirCorp and Space Adventures. Like them, it demonstrated how resourceful and determined private companies are redefining access to space.

During his undergraduate years at Brandeis University, Gold remembers writing a paper on the Strategic Defense Initiative as the Soviet Union fell. Little did he know that a decade and a half later he would find himself at a Russian nuclear missile base, launching a former nuclear weapon for a peaceful purpose. "As we like to say in Bigelow Aerospace: reducing the world's arsenal, one rocket at a time."

Gold's space pedigree extends back to his grandfather, who had been a radioman on the Apollo program. He remembers his grandfather's basement as being a "cross between the bridge of the Starship *Enterprise* and Frankenstein's lair."

While in law school at the University of Pennsylvania, Gold became a law clerk for NASA's Langley in 1996, which confirmed for him that he could pursue his love of the law and space at the same time. He got that opportunity while working for the DC law firm McGuire and Woods, where he represented space start-up companies, such as Gary Hudson's Rotary Rocket, and worked on spaceport initiatives in Virginia and Montana. That work brought him into contact with Robert Bigelow.

Bigelow would eventually retain Gold to do legal work, including spearheading the writing of a white paper dealing with the development of space and with NASA's problems competing with the private sector. To his surprise, they got a positive response from NASA. This started discussions on how NASA and Bigelow Aerospace might work together, and eventually led to Bigelow's $500 million commitment to invest in the development of expandable space habitat technology.

Following the signing of a memorandum of understanding in 2000, Bigelow visited NASA's Johnson Space Center in Texas. The meeting did not bode well for prospects of NASA collaboration. As Gold describes that first encounter, "The engineers were literally covering TransHab with blankets and laughing behind our back." However, things went better at the next meeting in early 2002. NASA's attitude toward Bigelow and Gold had unexpectedly warmed, which led to the development of the first Space Act Agreement (SAA) executed between NASA and a commercial company.

NASA's TransHab project was officially canceled in 2000. Gold remembers seeing a piece of TransHab fabric with "Rest in Peace" written on it. For Gold it represented a frustrating pattern of canceled NASA programs. "We saw that with X-33, orbital spaceplane, Venturestar, Alternative Access to Space, and Shuttle II. You need to change the acronym [NASA] to Never Any Sustainable Action," Gold notes. "They never complete what they start. You can't blame them, because it's the nature of the beast." The "beast" part of the equation being that the political and funding process changes every four to five years, too short a time for any space project to come to fruition.

However, dealing with NASA was much less of a problem than navigating through the red tape involved with launch certifications and export controls. With so much to be done in Washington, Bigelow hired Gold to head the new Bigelow Aerospace office in DC.

Gold's first order of business was to begin to work his way through the FAA payload approval process for the expandable habitat prototype demonstrator Genesis I. Payload certification would normally have been done by the launch provider, but since the Genesis I payload was unique, Bigelow Aerospace thought it prudent to get the approval itself. This was done through the FAA/AST office in DC.

Although Bigelow originally intended to use a U.S. launch provider, the cost factor led him to consider the use of Kosmotras, a joint Russian and Ukrainian company. Kosmotras was relatively new; it had launched only a few small satellites and had never launched a commercial payload prior to contracting with Bigelow Aerospace.

Kosmotras takes decommissioned Russian ICBMs, in this case SS18s, referred to as "Satan" by NATO, removes the warhead, replaces it with a commercial faring, and then launches it as a commercial space vehicle. Production and launches were conducted in Dnepropetrovsk, Ukraine, and in Yasny Space and Missile Complex, an active nuclear military base in the Orenburg region of Russia.

The "launch cost is somewhere near a factor of three times less than we would have paid with the nearest domestic equivalent," Gold explains. "Mr. Bigelow is a patriot. We never wanted to go outside the U.S. or specifically to Russia, but when you are looking at a dramatic difference in price and capability, you're really left with no choice. That's a further example of how America has lost the Space Race."

Unfortunately, the decision to use a foreign launch service brought the project under the realm of the State Department's export regulations known as ITAR, meant to control the export of defense-related articles and services.

"Second only to gravity, the International Traffic in Arms Regulations, or ITAR, had the greatest chance of preventing the spacecraft from ever leaving the ground," Gold recalls. Under ITAR, any space systems, including related equipment and hardware, were placed on the U.S. munitions list and were regulated under the Department of State's strict rules and procedures.

"We were shocked by the amount of bureaucracy and paperwork involved in getting the spacecraft to Russia and collaborating with our Ukrainian partners," Gold reflected. It took about a year to acquire a Technical Assistance Agreement (TAA), which allowed Bigelow Aerospace to engage in discussions with foreign entities. This process included having monitors from the Defense Technology Security Administration (DTSA) be present during any meetings with the foreign partners.

Two days before Gold and Bigelow were scheduled to travel to Russia for the first meeting with their Russian and Ukrainian partners, Gold received a call from the DTSA monitor informing him that the Ukrainians were not allowed in the technical discussions, even though the TAA had been in place for months.

It was too late to mount an appeal. They were put in the embarrassing position of having to kick their Ukrainian partners out of the room during discussions. Gold was more than a little disgusted with the process. "I got not one, but two DTSA monitors with me, breathing down my neck, monitoring every word that I say. To make matters worse, not only am I monitored like this, but I'm paying for it."

For Genesis I alone, Gold paid $220,000. It included a $130 per hour rate for each of the monitors, plus their travel and meal expenses. Gold is still more than a little bitter about the experience. "The KGB may have spied on the Russians but at least they had the good courtesy to do it for free."

For Gold it was the failure to effectively distinguish between benign, commercially available technologies and those with actual military relevance that was at the heart of the ITAR problem. There was nothing in the Genesis module that could not be purchased on the global market.

If you really want to see Gold's blood boil, just mention the Genesis I test stand. The test stand was basically a round piece of metal with four legs sticking up. "Flip it upside down, put on a white tablecloth, and it's indistinguishable from a metal coffee table. But as part of the TAA, I had to have two security guards guarding the table in Russia on a 24/7 basis! I can only imagine the national security implications of this table technology leaking to the Russians or the Ukrainians. They could be serving coffee on the damn thing." Gold had to file two general correspondence letters and spend several months to get permission not to guard the "coffee table."

Gold has no problem with the concept of export control. "There are absolutely technologies that require an extra degree of security, but it does not include coffee tables. All we're saying is that if you can purchase something in the open market or [you are] dealing with tech that is already obsolete or observably benign, it should not be covered under ITAR."

Along with the political adjustments that are required when cold war adversaries start to work together, commercial and social relationships evolve as well. The Genesis project was a case study in forging this new relationship.

Gold jokes that he's spent almost three years of his life in Siberia, only unlike the Russians, he knew he was coming back. His role in both the Genesis I and Genesis II launch campaigns grew from handling the legal and contract negotiations to becoming launch campaign manager on the ground in Yasny. During his Siberian stints, he stayed at the "Ritz Carlton of Siberia," a converted barracks built at the launch base especially for Bigelow staff. Gold was convinced that the hotel had one tube, which could deliver either Internet or hot water, but never both at the same time.

While at the base, Gold took to exposing his Russian and Ukrainian counterparts to bits of American culture. A die-hard Red Sox fan, he brought a load of baseball equipment and made sure to introduce the sport to the small Siberian town during a Memorial Day picnic. On other occasions he arranged a Fourth of July fireworks display on the Russian military base and rented the town's local theater to show the first *Star Wars* movie.

But Gold's experiences also exposed him to some of the Russians' peculiarities, such as their tendency for doing things their own way and inventing procedures on the fly. The day that the encapsulated Genesis I spacecraft was to move from the base to the silo, dubbed the "Big Joint Operations Day," was carefully planned. The craft was supposed to move in a caravan aboard a missile carrier with armed guards. But the Russians moved the schedule up one hour without informing Gold and his team.

The Russians naturally did not want Bigelow's staff to ride on the missile transporter, implementing their reasonable version of the Russian ITAR. But it was early in the season, and the spring melt had muddied the roads and deepened potholes. Worse, Genesis I was never designed to be transported horizontally. Gold grew increasingly alarmed when he thought there was no way of controlling the pace of the convoy and that the Russians were going to basically wing it. After working on the project for six years, he wasn't about to leave the operation to chance.

Just as they were about to move, Gold ran into the path of the convoy, wearing a Red Sox baseball hat and *Star Trek* T-shirt. He yelled "Stop!" and physically blocked the missile transporter. At that point he heard all the guns click. "This is probably the closest I ever came to being shot," he recalls. His Kosmotras representatives came to the rescue to explain what the crazy American was doing. Eventually, the solution was to put Gold's team in a van in front of the transporter to control the pace of the convoy as the van tested the road ahead.

Gold had also not anticipated the religious component of this partnership. On one occasion, while monitoring the launch silo with twenty-four-hour video cameras, which was a requirement for guarding the spacecraft under ITAR, the DTSA monitors saw a priest appear out of nowhere to bless the silo and the rocket. At first they thought it was a joke, but the man was seriously intent on giving his blessing.

Despite a few such incidents, Gold has only praise for the work of his Russian counterparts and their cooperation throughout both campaigns. He recalls having his calls returned from Moscow sometimes at 2:00 a.m. and more rapidly than from his subcontractors in Virginia. Even with all the difficulties involved, Gold notes with satisfaction, the Russians never once raised the price.

With the success of Genesis I, Bigelow decided to watch the Genesis II launch from Yasny for himself, and traveled to Russia together with Gold. On the day of the launch, the men had to indulge the good-luck ritual of playing a game of baseball. Gold had thrown the ball around for an hour before the launch of Genesis I and was determined to take the tradition a step further this time by actually playing the first ever baseball game in Yasny—the first ever baseball game on a Russian military base, for that matter—ahead of the Genesis II launch. He had shipped the baseball equipment over to Moscow but was very disappointed to hear that it had been held up in customs.

Fortunately, there were enough balls and gloves left over from the Genesis I campaign the year before to play ball. Miraculously, one of the security guards even produced a baseball bat. Bigelow personnel and the DTSA monitors formed one team. The opposition consisted of twelve- to sixteen-year-old girls recruited from the town's ballet school. The weather was not cooperating, so General Andreyev, the base director, offered the use of the Assembly Integration and Test Building, where the spacecraft was prepped. Bob Bigelow smacked a double, and at the end of the game he donated $1,000 to the school, much to the delight of everyone.

The Genesis II launch on 28 June 2007 did not go as smoothly as the lift-off of Genesis I had eleven months earlier. It normally takes fifteen minutes from launch to rocket separation, but just as the moment of separation was to occur, they lost telemetry from the rocket. "Did this happen during the Genesis I launch?" Bigelow asked Gold. "No, sir, it did not," Gold nervously replied.

Finally, word came in over mission control's phone lines that the space-craft had successfully separated. Apparently the ground station in Irkutsk had lost connection with Yasny but not with the rocket, and had been unable to transmit the information. Everyone erupted in joy and relief with the good news. The Russians had placed Genesis II within one hundred meters of its planned parking orbit. "I can't say enough how much working with the Russians really meant," said Gold, "taking a weapon of war, transforming it into a tool of peaceful commerce. It was horrendous looking back that we had these weapons pointing at each other. We're not just trying to develop a better technology, we're trying to develop a better future. And I'm very glad the Genesis missions were able to play that role."

Today, there are two privately owned, Bigelow mini-stations in low Earth orbit, downloading live images of the Earth and their internal modules, something that a few years ago was deemed impossible. Genesis II pioneered the "Fly Your Stuff" program carrying hundreds of small personal memorabilia up into space for just $300. After a few months in orbit, they had successfully photographed every single article flown, through cameras mounted inside the module.

The next phase for Bigelow Aerospace is building the habitat Sundancer, which Gold describes as a "quantum leap from autonomous to actual crewed vehicles." Once Sundancer is launched and flight-proven, Bigelow plans to launch a node bus combination that will serve as the base for connecting multiple Sundancers and maintaining the cluster of habitats in designated orbit.

Eventually, Bigelow will launch its "Model T" module, called the BA-330, which is already under development in Las Vegas. The BA-330 is a single habitat that can provide 230 cubic meters of usable volume—about twice the size of Sundancer. It doesn't take long to realize that such a combination of modules could pass the volume of the ISS pretty quickly.

However, Bigelow Aerospace is well aware that one critical component is still missing from its business plan. "So we can get Sundancer up there, but once we do, what happens next?" Gold asks. "With cargo there are many options: the Russian Progress, Europe's Autonomous Transfer Vehicle (ATV), and soon Japan's H-II Transfer Vehicle. But with crew, there's literally no option. Transportation is really our Achilles heel."

This is why Bigelow, along with the rest of the private space community, is monitoring Elon Musk's progress in developing a low-cost transportation system, and also looking at other potential options. Bigelow has signed a memorandum of understanding with United Launch Alliance to look at evolved expendable launch vehicle options transport and is open to other foreign launch possibilities if they become viable.

Once Bigelow's huge inflatable structures are in place, it doesn't take any great stretch of the imagination to conjure visions of communities living and working in space, or to see a reliable, safe, inexpensive SpaceX transport system ferrying humans back and forth to these bases.

According to the forecasts of Gerard O'Neill, we are already behind the development curve for opening up space. But thanks to the efforts of some individuals who were inspired by NASA's early space efforts and by O'Neill's vision of our space future, the movement seems back on track. Many of the individuals involved in private space share O'Neill's grand vision of the human necessity for access to space and the enormous impact it will have on our lives.

In her flight blog, private space explorer Anousheh Ansari spoke of the power of looking down on Earth and feeling connected to the rest of humanity. Peter Diamandis talks about our "moral imperative" to open up the space frontier to achieve "planetary redundancy, to back up the biosphere," It is not unusual to hear Virgin Galactic's Will Whitehorn speak in O'Neillian terms of space as mankind's destiny. "Man has to go to space. It's not about the science; it's about the principle. We have to explore or we lose our sense of civilization."

O'Neill's vision of humanity's future in space was held captive to the ability of new technologies to deliver on those dreams. Likewise, the advent of commercial spaceflight has been running on dreams, working on the technology, and waiting for the dreams to become reality.

By the beginning of 2010, as this book headed to press, both WhiteKnightTwo and SpaceShipTwo had made their long-awaited debuts. On 27 July 2009 WhiteKnightTwo flew long distance from Mojave Spaceport for its first public appearance before a huge crowd at the Experimental Aircraft Association's AirVenture air show in Oshkosh, Wisconsin.

SpaceShipTwo received its unveiling on the cold and blustery evening of 7 December 2009 at Mojave Air and Spaceport before some eight hundred invited guests and reporters. The event bore the stamp of a new marketing vision, something more akin to Hollywood than to Mojave: klieg lights, young starlets, free champagne, Scaled Composites engineers being introduced as "rock stars," and Virgin Galactic's future astronauts dressed in black designer Puma jackets with "ss2 Unveil" logos. It was a glitzy wrap-up to the past six years.

Back in 2003, at the unveiling of SpaceShipOne, Rutan had been quick to point out that the event was not about "dreams, predictions, or mock-ups." It was instead about actual flight hardware and a complete manned space program. The 7 December event was also about hardware, but there was no shortage of dreams and the selling of dreams.

With the winds building to gale force, the crowd stepped outside the large reception tent. On cue, the entire complex went pitch-dark, music swelled, and one by one spotlights snapped on to illuminate the runway and the three-story-tall display with icons depicting the history of flight. Then from the other end of the runway WhiteKnightTwo taxied forward, bathed in purple light and carrying beneath its twin fuselage the star of the show—Rutan's new SpaceShipTwo.

An enthusiastic cheer went up as the twin aircraft stopped in front of the crowd. At sixty feet in length, SpaceShipTwo was twice the size of its predecessor. Designed to accommodate eight people per flight, the spacecraft will reach altitudes between eighty-four and eighty-seven miles and provide four and a half minutes of weightlessness.

Here was the hardware. No longer prototypes or experimental aircraft, these were the vehicles that will actually take paying customers to space. In two years or three, when flight testing is complete, Scaled Composites will fill out Virgin Galactic's fleet of commercial, passenger-carrying spaceships with four more SpaceShipTwos. Rutan is predicting that forty or fifty spacecraft will eventually be needed to meet market demand.

In the pomp of this event, it was easy to imagine a tidal shift in access to space, as if the whole private space industry had just stepped forward to stand on a more even footing with NASA. If current schedules stand, in 2010 NASA is expected to retire its three space shuttles and rely on SpaceX's Falcon 9 for access to the ISS. Although its life may be extended, current plans are to deorbit the ISS in 2016, when Bigelow Aerospace may begin parking private stations in Earth orbit and beyond.

On that December evening, as SpaceShipTwo stood bathed in dramatic spotlight, decades of dreams for private access to space took on substance. Along with Burt Rutan, dozens of other individuals and companies that have been working to push the boundaries of space access now stand poised to stake their claim on the cosmos.

Sources

Books

Atkinson, Joseph D., and Jay M. Shafritz. *The Real Stuff: A History of NASA's Astronaut Recruitment Program*. New York: Praeger, 1985.

Belfiore, Michael. *Rocketeers: How a Visionary Band of Business Leaders, Engineers, and Pilots Is Boldly Privatizing Space*. New York: Smithsonian Books, 2007.

Brand, Stewart, ed. *Space Colonies*. New York: Penguin Books, 1977.

Branson, Richard. *Losing My Virginity: The Autobiography*. New York: Virgin Books, 2007.

———. *Screw It, Let's Do It: Expanded Edition*. New York: Virgin Books, 2008.

Burgess, Colin. *Teacher in Space: Christa McAuliffe and the Challenger Legacy*. Lincoln: University of Nebraska Press, 2000.

Corrigan, Grace George. *A Journal for Christa: Christa McAuliffe, Teacher in Space*. Lincoln: University of Nebraska Press, 1993.

Dethloff, Henry C. *Suddenly Tomorrow Came . . . A History of the Johnson Space Center*. Washington DC: NASA, 1993.

Feoktistov, Konstantin. *The Trajectory of Life*. Moscow: Vagrius, 2000.

Flight of Spaceship Voskhod. Moscow: Novosti, 1964.

Garber, Stephen J., ed. *Looking Backward, Looking Forward: Forty Years of U.S. Human Spaceflight Symposium*. Washington DC: NASA, 2002.

Hall, Rex, and David J. Shayler. *The Rocket Men, Vostok & Voskhod: The First Soviet Manned Spaceflights*. Chichester, UK: Springer-Praxis, 2001.

Hall, Rex D., David J. Shayler, and Bert Vis. *Russia's Cosmonauts*. Chichester, UK: Praxis, 2005.

Harford, James. *Korolev*. New York: John Wiley & Sons, 1997.

Hudgins, Edward, ed. *Space: The Free Market Frontier.* Washington DC: Cato Institute, 2002.

Johnson, R. D., and C. Holbrow, eds. *Space Settlements: A Design Study.* Washington DC: NASA, 1976.

Kamanin, Nikolai. *"I Feel Sorry for Our Guys": General N. Kamanin's Space Diaries.* Moscow, 1993 (NASATT-21658, 1994).

Klerkx, Greg. *Lost in Space: The Fall of NASA and the Dream of a New Space Age.* New York: Pantheon Books, 2004.

Lascarides, Effie. *Apollo's Legacy: The Hellenic Torch in America at the Dawn of the New Millennium.* Brookline NY: Hellenic College Press, 2000.

Lindbergh, Charles A. *The Spirit of St. Louis.* New York: Scribner, 1953.

Linehan, Dan. *SpaceShipOne: An Illustrated History.* Minneapolis: Zenith Press, 2008.

Mailer, Norman. *Of a Fire on the Moon.* Boston: Little, Brown, 1970.

Maryniak, Gregg. "When Will We See a Golden Age of Space Flight?" In *Space: The Free Market Frontier,* ed. Edward Hudgins. Washington DC: Cato Institute, 2002.

Matson, Wayne R., ed. *Cosmonautics: A Colorful History.* Washington DC: Cosmos Books, 1994.

Michaud, Michael A. G. *Reaching for the High Frontier.* New York: Praeger, 1986.

Mullane, Mike. *Riding Rockets: The Outrageous Tales of a Space Shuttle Pilot.* New York: Scribners, 2006.

O'Neill, Gerard K. *The High Frontier: Human Colonies in Space,* 3rd ed. Burlington, Ontario: Collector's Guide Publishing, 2000 [1977].

Regis, Ed. *Great Mambo Chicken & the Transhuman Condition: Science Slightly over the Edge.* New York: Addison-Wesley Publishing Co., 1990.

Richelson, Jeffrey T. *America's Secret Eyes in Space: The U.S. KEYHOLE Spy Satellite Program.* New York: Harper & Row, 1990.

Sharman, Helen, and Christopher Priest. *Seize the Day.* London: Victor Gallancz, 1993.

Shayler, David J., and Colin Burgess. *NASA's Scientist-Astronauts.* Chichester, UK: Springer-Praxis, 2007.

Siddiqi, Asif A. *Challenge to Apollo: The Soviet Union and the Space Race, 1945–1974.* Washington DC: NASA, 2000.

Slayton, Donald K. "Deke," with Michael Cassutt. *Deke: U.S. Manned Space:*

From Mercury to the Shuttle. New York: Tom Doherty Associates, 1994.

Stine, G. Harry. *Halfway to Anywhere: Achieving America's Destiny in Space.* New York: Smithsonian Books, 1996.

Thomas, Shirley. *Men of Space: Profiles of the Leaders in Space Research, Development, and Exploration,* vol. 2. New York: Chilton Co., 1961.

Weil, Elizabeth. *They All Laughed at Christopher Columbus: An Incurable Dreamer Builds the First Civilian Spaceship.* New York: Bantam Books, 2002.

Wilbur, Ted. *Space and the United States Navy.* Washington DC: Chief of Naval Operations, 1970.

Periodicals and Online Articles

Aaron, Kenneth. "Space Tourists: XCOR's Suborbital Rocket Could be the Ultimate Thrill in Adventure Travel." *Cornell Engineering Magazine,* May 2003.

Academy of Achievement. "Biography: Jeff Bezos." http://www.achievement.org/autodoc/printmember/bezobio-1.

Adams, Eric. "The New Right Stuff." *Popular Science,* November 2004.

Artemis Project. "Where the NSS Local Chapters Came From." http://www.asi.org/adb/06/01/nss-chapters-origins.html.

Asimov, Isaac. "The Cruise and I." *The Magazine of Fantasy and Science Fiction,* July 1973.

————. "The Next Frontier?" *National Geographic,* July 1976.

Belfiore, Michael. "Live Coverage: Virgin Galactic Unveils SpaceShipTwo." *Wired* Blog Network, 23 January 2008. http://blog.wired.com/wiredscience/2008/01/spaceshiptwo-un.html.

Bohlen, Celestine. "Russian Astronauts Plead with Public to Save the Mir." *New York Times,* 14 June 1999.

Bonné, Jon. "Private Manned Space Plane Unveiled." *MSNBC.com,* 18 April 2003. http://www.msnbc.msn.com/id/3077811/ns/technology_and_science-space/.

Borel, Brooke. "The Trip Was So Nice, He's Going up Twice: Q&A with a Billionaire Space Tourist." *PopSci.com,* 13 February 2009. http://www.popsci.com/military-aviation-amp-space/article/2009-02/trip-was-so-nice-he%E2%80%99s-going-twice-qampa-billionaire-space-tourist.

Boyle, Alan. "Millionaire Resumes Space Training." *MSNBC.com*, 25 May 2005. http://www.msnbc.msn.com/id/7978430/.

———. "Russians Block Space Passenger's Trip." *MSNBC.com*, 23 June 2004. http://www.msnbc.msn.com/id/5278674/ns/technology_and _science-space/.

———. "Space Passenger Rides Out Highs and Lows." *MSNBC.com*, 1 October 2005. http://www.msnbc.msn.com/id/9313347/.

Brandt-Erichson, David. "The L5 Society." *Ad Astra*, November–December 1994.

Brekke, Dan. "Who Needs NASA: Do-It-Yourself Astropreneurs Are Bucking the System to Put a Schmo in Orbit." *Wired*, January 2000.

Buckley, Tom. "Caribbean Cruise Attempts to Seek Meaning of Apollo." *New York Times*, 12 December 1972.

"California Company Shoots for Cheap Ride to Space." *CNN.com*, 10 January 2002. http://articles.cnn.com/2002-01-10/tech/ez.rocket_1_xcor -aerospace-rocket-flight?_s=PM:TECH.

Cassutt, Michael. "Citizen in Space." *Space.com*, 28 March 2001. http:// www.space.com.

Chafkin, Max. "Entrepreneur of the Year: Elon Musk." *Inc.*, December 2007.

Chang, Alicia. "Pioneering Space Tourism Company Depends on Clients to Create Buzz." *USA Today*, 30 July 2006.

"Civilian Space Travel: Got Cash?" *Wired*, 27 March 1998.

Clash, James. "Space Cowboy." *Forbes.com*, 9 May 2005. http://www.forbes .com/forbes/2005/0509/058.html.

Coppinger, Rob. "Whatever Happened to Blue Origin?" *Hyperbola*, 8 December 2008. http://www.flightglobal.com.

Corsello, Andrew. "The Believer." *Mens.Style.com*. http://mens.style.com/ gq/features/landing?id-content_8037.

Covault, Craig. "Bigelow Reveals Space Business Plan." *Aviation Week*, 6 April 2007.

Cowling, Keith. "Earth's First Self-Financed Astronaut." *SpaceRef.com*, 10 December 2000. http://spaceref.com/news/viewnews.html?id=263.

———. "Mr. Tito Comes to Washington—Part 2." *SpaceRef.com*, 23 May 2001. http://www.spaceref.com/news/viewnews.html?id=349.

DarkSyde. "Interview with an Astronaut." *Dailykos.com*, 6 September 2009.

http://www.dailykos.com/story/2009/9/6/776871/-Interview-with
-an-Astronaut.

Da Silva, Wilson. "Children of Apollo." *Cosmos*, December 2006.

David, Leonard. "Beyond Tito: Space Travelers Wanted." *Space.com*, 1 May 2001. http://www.space.com/businesstechnology/technology/tito _next_step_010501-1.html.

———. "NASA Chief Remains Miffed Over Tito Launch: 'Space Is Not About Egos.'" *Space.com*, 28 April 2001. http://www.space.com/ missionlaunches/missions/tito_pm_010428.html.

———. "Spaceport Taking Shape in Mojave Desert." *MSNBC.com*, 17 June 2004. http://www.musnbc.nsn.com/id/5051823/1/displaymode/1098/.

Davisson, Budd. "The EZ Rocket: No Really, We're Not Kidding!" *Sport Aviation*, June 2003.

De Selding, Peter B. "ESA Space Station Chief Says Nyet to Tito's Visit." *Space .com*, 2 February 2001. http://www.space.com/missionlaunches/ launches/esa_contra_tito_010202.html.

Deutschman, Alan. "The Gonzo Way of Branding." *Fast Company*, October 2004. http://www.fastcompany.com/magazine/87/branson.html.

Diamandis, Peter, and Gregg E. Maryniak. "Why the X Prize?" *Ad Astra*, May–June 1998.

Drexler, Eric. "Have We Changed Our Goals?" *L5 News*, October 1983.

Duignan-Cabrera, Anthony. "Dennis Tito Says It's 'Highly Likely' He Will Go to the ISS in April 2001." *Space.com*, 16 November 2000. http:// www.space.com/news/spacestation/tito_to_iss_001116-2.html.

Dyson, Freeman. "Gerard Kitchen O'Neill." *Physics Today*, February 1993.

"Earth to Gary." *Forbes.com*. http://www.forbes.com/forbes/1999/0705/64 01140a_print.html.

"Ehricke." *Encyclopedia Astronautica*. www.astronautix.com.

Ehricke, Krafft A. "The Extraterrestrial Imperative." *Air University Review*, January–February 1978.

"Fly Me to the Moon." *Retro Future*. http://www.retrofuture.com.

Foust, Jeff. "Burt Rutan, in His Own Words." *Space Review*, 25 October 2004. http://www.thespacereview.com/article/255/1.

France, David. "Houston, We Have a Problem." *Newsweek*, 20 March 2001.

———. "One Real Space Cowboy." *Newsweek*, 2 April 2001.

Freedman, David H. "Burt Rutan: Entrepreneur of the Year." *Inc.*, January 2005.

"Free Enterprise Space Shot." *Time*, 29 June 1981.

Ganoe, William. "Rockets from the Sea." *Ad Astra*, July–August 1990.

Garrison, Peter. "Top Pencil." *Aviation History*, November 2009.

Gatlin, Allison. "Company's Dreams Skyrocket: Private Firm Tests Civilian Spacecraft." *Antelope Valley Press*, 24 November 2001.

Geist, William E. "Sky Is the Limit for Citizen's Backyard." *New York Times*, 12 April 1981.

Gold, Mike. "Lost in Space: A Practitioner's First-Hand Perspective on Reforming the U.S.'s Obsolete, Arrogant, and Counterproductive Export Control Regime for Space-Related Systems and Technologies." *Journal of Space Law* 34, no. 1 (2008).

Goudarzi, Sara. "Interview with Anousheh Ansari, the First Female Space Tourist." *Space.com*, 15 September 2006. http://www.space.com/missionlaunches/060915_ansari_qna.html.

Gross, Jerry. "Arfons' Water Bomb." *Hot Rod*, November 1966.

Halvorson, Todd. "NASA, Russia to Discuss Sending Tourists to International Space Station." *Space.com*, 26 January 2001. http://www.space.com/news/russia_usa_tourists_010126.html.

———. "NASA Continues Protesting Space Joyride of Dennis Tito." *Spaceflight Now*, 20 March 2001. http://spaceflightnow.com/news/no103/20tito/.

———. "Tito: NASA Not an Issue Regarding ISS Trip." *Space.com*, 1 February 2001. http://www.space.com/missionlaunches/tito_interview_010201.html.

Henderson, Diedtra. "Personal Space Trip: $98,000 Idea." *Seattle Times*, 21 October 1997.

Henson, Keith. "Bulletin from the Moon Treaty Front." *L5 News*, January 1980.

Hsu, Jeremy. "British Spaceplane Gets Boost." *Space.com*, 11 March 2009. http://www.space.com/businesstechnology/090311-tw-space-plane.html.

Hudson, Gary C. "History of the Phoenix VTOL SSTO and Recent Developments in Single-Stage Launch Systems." *SpaceFuture.com*, November 1991. http://www.spacefuture.com/archive/history_of_the_phoe

nix_vtol_ssto_and_recent_developments_in_single_stage_launch
_systems.shtml.

———. "Insanely Great? Or Just Plain Insane?" *Wired*, May 1996.

———. "Interview with Gary Hudson: Space Entrepreneur & RLV Designer." *HobbySpace.com*, June 2003. http://www.hobbyspace.com/ AAdmin/archive/Interviews/Systems/GaryHudson.html.

Hum, Peter. "Rocket Man." *Ottawa Citizen*, 28 September 2006.

Juran, Ken. "Space Tourist Greg Olsen on Riding Soyuz, Orbiting Earth and Visiting the ISS." *Popular Mechanics*, October 2005.

"Kamanin Diaries." *Encyclopedia Astronautica*. http://www.astronautix.com /articles/kamaries.htm.

Kanellos, Michael. "Newsmaker: Elon Musk on Rockets, Sports Cars, and Solar Power." *CNET News*, 15 February 2008.

Karash, Yuri. "Even if MirCorp Finds Funds, Will It Be Too Late?" *Space. com*, 19 October 2000. http://www.space.com/news/spacestation/ mir_officials_001019.html.

———. "MirCorp President Jeffrey Manber Argued in a Letter to Russian President Putin That He Must Help Save the Mir Space Station." *Space.com*, 27 October 2000. http://www.space.com/news/space shuttles/mir_letter_001027.html.

———. "Mir Movie Will Boost Morale." *Space.com*, 1 February 2000. http:// www.space.com/businesstechnology/business/mir_movie_sidebar1 _000201.html.

Kluger, Jeffrey. "NASA Goes Hollywood?" *Time*, 1 October 2000.

Law, Glenn. "A Vacation in Space: The Aerospace Corporation's Support of Spaceport Development." *Crosslink*, Winter 2008.

Leary, Warren E. "Millionaire Hopes to Be First Tourist in Space." *New York Times*, 20 June 2000.

"L5 in Congress." *L5 News*, April 1977.

Lindsey, Clark S. "Climbing a Commercial Stairway to Space: A Plausible Timeline? Version 2009." *HobbySpace.com*, 2 February 2009. http://www.hobbyspace.com/AAdmin/Archive/SpecialTopics/to Space/TimeLine.html.

Little, Geoffrey. "Mr. B's Big Plan: Robert Bigelow Has Put Two Mini-Space Stations in Orbit. Now Comes the Hard Part." *Air & Space Magazine*, 1 January 2008.

Malik, Tariq. "Space Adventures Teams with XCOR Aerospace to Develop Sub-orbital Vehicle." *Space.com*, 22 July 2002. http://www.space .com/missionlaunches/space_tourism_020722.html.

Meisler, Andy. "At Home With: Burt Rutan; Slipping the Bonds of Earth and Sky." *New York Times*, 3 August 1995.

Miller, Terry. "Walt Arfons' Steam Dragster." *American Rodding*, June 1967.

Motta, Mary. "The Men behind Mir's Financial Rescue." *Space.com*, 7 April 2000. http://www.space.com/news/spacestation/anderson_profile _000407.html.

Nagy, Kim, and Joy Stocke, "Reaching for the Stars: An Interview with Greg Olsen — Scientist, Entrepreneur, and Space Traveler." *Wild River Review*, August 2009. http://www.wildriverreview.com/spot light_olsen.php.

Oberg, James. "Can Lance Sing His Way into Space?" *Spacedaily.com*, 28 August 2002. http://www.spacedaily.com/news/tourism-02m.html.

———. "Russia's Space Program Running on Empty: Part I." *Spectrum Magazine*, December 1995.

O'Neill, Gerard K. "The Colonization of Space." *Physics Today*, September 1974.

———. Interview by Keith Henson and Carolyn Henson. *L5 News*, March 1977.

Owens, Jim. "Take My Rocket, Please." *SpaceDaily.com*, 5 November 1999. http://www.spacedaily.com/news/launchers-99s.html.

Reiss, Spencer. "Rocket Man." *Wired*, January 2005.

Robinson, Ann Elizabeth. "Space Industrialization: Captain Frietag of NASA." *L5 News*, March 1976.

"Rotary Rocket." *Absolute Astronomy.com*. http://www.absoluteastronomy. com/topics/Rotary_Rocket.

Rothenberg, Randall. "Launching by Soviets Succeeds, but Advertising Plan Fails." *New York Times*, 21 May 1991.

Sanger, David. "A Japanese Innovation: The Space Antihero." *New York Times*, 8 December 1990.

Scaled Composites. "Tier One — Private Manned Space Flight." http://www .scaled.com/projects/tierone/message.htm.

Schmidt, William E. "London Journal; First Briton in Space, but Barely in Fame's Orbit." *New York Times*, 25 May 1991.

Schwartz, John. "At One Point, 'I was Deathly Afraid,' New Space Visitor Admits." *New York Times*, 23 June 2004.

———. "Thrillionaires: The New Space Capitalists." *New York Times*, 14 June 2005.

Siegler, Paul. "Be Your Own Astronaut." *15 News*, April 1978.

Snelson, Robin. "X-Prize Losers: Still in the Race, Not Doing Anything, or Too Sexy for the X Cup?" *Space Review*, 26 September 2005.

"Soviet Reporter Tells of Takeoff." *New York Times*, 13 October 1964.

Stone, Brad. "Amazon Enters the Space Race." *Wired*, July 2003.

Sullivan, Walter. "Princeton Conference Makes Detailed Assessment of Problems of Establishing a Colony in Space." *New York Times*, 12 May 1975.

———. "Proposals for Colonies in Space Hailed as Feasible by Scientists." *New York Times*, 13 May 1974.

———. "Specialists in Space." *New York Times*, 13 October 1964.

Sweetman, Bill. "Burt Rutan Builds Your Ride to Space." *Popular Science*, July 2003.

Tanner, Henry. "Astronauts on TV." *New York Times*, 13 October 1964.

Taylor, Chris. "The Sky's the Limit." *Time*, 29 November 2004.

Tedeschi, Diane. "Sweet Success." *Air & Space Magazine*, February–March 2006.

Trevisan, Matthew. "Three, Two, One — Commence Bragging." *Globe and Mail*, 8 July 2007.

Truax, Robert C. "Shuttles: What Price Elegance?" *Astronautics and Aeronautics*, June 1970.

"200 on Pan Am Waiting List Aiming for Moon." *New York Times*, 9 January 1969.

Tyler, Patricia. "Space Tourist, Back From 'Paradise,' Lands on Steppes." *New York Times*, 7 May 2001.

"Virgin Galactic: Work in Progress." *Hyperbola*, 26 March 2008. http://www.flightglobal.com/blogs/hyperbola/2008/03/virgin-galactic-work-in-progress-1.ntml#more.

Von Braun, Wernher. "Space Pioneer Reflects on Apollo's Achievements." *New York Times*, 3 December 1972.

"Voskhod 1." *Encyclopedia Astronautica*. http://www.astronautix.com/flights/voskhod1.htm.

Wald, Matthew L. "2 Space Novices with a Love of Knowledge: Christa McAuliffe." *New York Times*, 10 February 1986.

Weil, Elizabeth. "American Megamillionaire Gets Russki Space Heap." *New York Times*, 23 July 2000.

"Where's Jim Tighe?" *Design News*, 7 March 2005. http://goliath.ecnext .com/coms2/gi_0198-209948/Space-cadet-Jim-Tighe-was.html.

Whittle, Lisa. "Falling Down in the Race for Space." *New Scientist*, 10 August 1991.

Wilford, John Noble. "Risks Spelled Out, Shuttle 'Survivor' Says." *New York Times*, 21 April 1986.

Wilson, Anton Wilson. "New Ames Space Programs." *L5 News*, March 1976.

Windrem, Robert. "Inside the Mind of a Space Tycoon." *MSNBC.com*, 6 May 2001. http://www.msnbc.msn.com/id/3077967/.

"XFlight." *The Shocker: The WSU Alumni Magazine Online Edition*, 29 August 2008. http://webs.wichita.edu/dt/shockermag/show/features .asp?_s=137.

Interviews and Personal Communications

Adams, Craig. Interview by Chris Dubbs (telephone), 26 September 2007.

Anderson, Eric. Interview by Emeline Paat-Dahlstrom, Arlington VA, 12 March 2008. E-mail correspondence, 11 November 2008.

Anderson, Walt. Interview, *The Space Show*, 25 April 2004.

Ansari, Anousheh. Interview by Emeline Paat-Dahlstrom (telephone), 18 June 2008.

Binnie, Brian. Interview by Emeline Paat-Dahlstrom (telephone), 4 September 2008.

Bolden, Charles F. Interview by Sandra Johnson, 6 January 2004. NASA Johnson Space Center Oral History Project, Oral History transcript. Houston TX: NASA Johnson Space Center.

Bova, Ben. Interview by Chris Dubbs, Dallas TX, 25 May 2007.

Carmack, John. Interview by Emeline Paat-Dahlstrom, Washington DC, 28 May 2008.

Clamon, Randy. Interview by Chris Dubbs (telephone), 12 March 2007.

Collier, Ian. Interview by Emeline Paat-Dahlstrom (telephone), 15 October 2008.

Davidian, Ken. Interview by Emeline Paat-Dahlstrom (telephone), 30 June 2008 and 1 July 2008. Arlington VA, 7 February 2009.

Davidian, Ken and Gretchen. Interview by Emeline Paat-Dahlstrom and Eric Dahlstrom, Arlington VA, 7 May 2008.

Davidson, James. E-mail correspondence with Chris Dubbs, 1 September 2008.

Davis, Don. E-mail correspondence with Chris Dubbs, 14 April 2007.

———. Interview by Chris Dubbs (telephone), 10 April 2007 and 13 June 2007.

Davis, Hugh. Interview by Chris Dubbs (telephone), 30 May 2007 and 5 June 2007.

Diamandis, Peter. Interview by Emeline Paat-Dahlstrom and Eric Dahlstrom (telephone), 6 June 2008.

Dunstan, James. Interview by Chris Dubbs (telephone), 17 January 2009.

Foust, Jeff. Interview by Emeline Paat-Dahlstrom and Eric Dahlstrom (telephone), 27 March 2009.

Fuqua, Don. Interview by Catherine Harwood, 11 August 1999. NASA Johnson Space Center Oral History Project, Oral History 2 transcript. Houston TX: NASA Johnson Space Center.

Garriott, Owen K. Interview by Kevin M. Rusnak, 6 November 2000. NASA Johnson Space Center Oral History Project, Oral History transcript. Houston TX: NASA Johnson Space Center.

Gold, Mike. Interview by Emeline Paat-Dahlstrom and Eric Dahlstrom (telephone), 10 February 2009 and 12 February 2009.

Hannah, Eric. Interview by Chris Dubbs (telephone), 27 March 2007.

Hidalgo, Loretta. Interview by Emeline Paat-Dahlstrom and Eric Dahlstrom (telephone), 6 November 2008.

Hopkins, Mark. Interview by Chris Dubbs, Dallas TX, 25 May 2007.

Hudson, Gary. Interview by Emeline Paat-Dahlstrom (e-mail), 12 May 2009.

Johnson, Richard. Interview by Chris Dubbs (telephone), 5 April 2007.

Kobrick, Ryan. Interview by Emeline Paat-Dahlstrom (telephone), 9 July 2008 and 12 February 2009.

Ladwig, Alan. Interview by Chris Dubbs, Washington DC, 30 May 2008 and (telephone) 22 July 2008.

Landeene, Steve. Interview by Emeline Paat-Dahlstrom and Eric Dahlstrom, Arlington VA, 5 February 2009.

Lauer, Chuck. Interview by Emeline Paat-Dahlstrom (telephone), 18 June 2008.

Leestma, David C. Interview by Jennifer Ross-Nazzal, 26 November 2002. NASA Johnson Space Center Oral History Project, Oral History transcript. Houston TX: NASA Johnson Space Center.

Lentsch, Aron. Interview by Emeline Paat-Dahlstrom and Eric Dahlstrom (telephone), 6 March 2009.

Malewicki, Dale. Interview by Chris Dubbs (telephone), 2 January 2008.

Manber, Jeffrey. Interview by Chris Dubbs (telephone), 27 August 2008.

Maryniak, Gregg. Interview by Emeline Paat-Dahlstrom and Eric Dahlstrom (telephone), 24 June 2008.

Messier, Doug. Interview by Emeline Paat-Dahlstrom and Eric Dahlstrom (telephone), 17 March 2009. Mountain View CA, 7 August 2009.

Moltzan, John. Interview by Emeline Paat-Dahlstrom, Washington DC, 22 September 2008.

Mullane, Richard D. Interview by Rebecca Wright, 24 January 2003. NASA Johnson Space Center Oral History Project, Oral History transcript. Houston TX: NASA Johnson Space Center.

Musk, Elon. Interview by Emeline Paat-Dahlstrom, Washington DC, 29 May 2008.

Nagel, Steven R. Interview by Jennifer Ross-Nagel, 20 December 2002. NASA Johnson Space Center Oral History Project, Oral History transcript. Houston TX: NASA Johnson Space Center.

Oelerich, John. Interview by Chris Dubbs (telephone), 10 January 2008.

Onuki, Misuzu. Interview by Emeline Paat-Dahlstrom and Eric Dahlstrom (telephone), 9 March 2009.

Owens, Brook. Interview by Emeline Paat-Dahlstrom, 18 February 2009.

Pearlman, Robert. Interview by Emeline Paat-Dahlstrom (video conference), 5 March 2008.

Peters, Fell. Interview by Chris Dubbs (telephone), 11 December 2007.

Schmitt, Harrison H. "Jack." Interview by Carol Butler, 14 July 1999. NASA Johnson Space Center Oral History Project, Oral History transcript. Houston TX: NASA Johnson Space Center.

Slater, Dan. Interview by Chris Dubbs (telephone), 8 June 2007.

Smith, Patti Grace. Interview by Emeline Paat-Dahlstrom and Eric Dahlstrom, Washington DC, 26 February 2009.

Sternback, Rick. Interview by Chris Dubbs (telephone), 13 April 2007.

Stinemetze, Matt. Interview by Emeline Paat-Dahlstrom (telephone), 16 September 2008.

Truax, Scott. Interview by Chris Dubbs (telephone), 29 June 2007.

Tumlinson, Rick. Interview, *The Space Show*, 30 October 2002.

van Hoften, James D. Interview by Jennifer Ross-Nazzal, 5 December 2007. NASA Johnson Space Center Oral History Project, Oral History transcript. Houston TX: NASA Johnson Space Center.

Walker, Charles D. Interview by Jennifer Ross-Nazzal, 19 November 2004, 17 March 2005. NASA Johnson Space Center Oral History Project, Oral History transcript. Houston TX: NASA Johnson Space Center.

Whitehorn, Will. Interview by Emeline Paat-Dahlstrom, Arlington VA, 6 February 2009.

Wimmer, Per. E-mail correspondence to Emeline Paat-Dahlstrom, 9 August 2009.

———. Interview by Emeline Paat-Dahlstrom, Washington DC, 30 May 2008.

Other Sources

"A Conversation with Amazon.com CEO Jeff Bezos." *Charlie Rose Show*, 19 November 2007.

Advent Launch Services. Web site. http://www.adventlaunchservices.com.

Ansari, Anousheh. "Anousheh Ansari Space Blog." September–October 2006. http://spaceblog.xprize.org/.

Ansari, Anousheh, Greg Olsen, and Eric Anderson. International Space Development Conference presentation, Washington DC, 31 May 2008.

Antunano, Melchor. "Promoting Medical Safety in Manned Commercial Spaceflights: A Key Factor for the Success of This Emerging Market." Paper presented at Eleventh Annual FAA Commercial Space Transportation Conference, Washington DC, 5 February 2008.

Armadillo Aerospace. Web site. http://www.armadilloaerospace.com.

Bigelow Aerospace. Web site. http://www.bigelowaerospace.com.

Black Sky: The Race for Space. DVD. Silver Spring MD: Discovery Channel, 2003.

Black Sky: Winning the X-Prize. DVD. Silver Spring MD: Discovery Channel, 2005.

Blue Origin. Web site. http://public.blueorigin.com/nsresearch.html.

Bronsz, Tom. Response to online forum topic, "Whatever happened to the ROTON?" *SpaceFellowship.com*, 11 October 2004.

Canadian Arrow. Web site. http://www.canadianarrow.com.

Collins, Patrick. "Prospects for Space Tourism in Japan." First International Symposium on Private Human Access to Space, Arcachon, France, 28–30 May 2008.

Compton, W. David. *Where No Man Has Gone Before*. NASA Special Publication-4214. Washington DC: NASA, 1989.

Conant, Eve. "Russia / Space." Voice of America broadcast, 20 January 2000. http://www.fas.org/news/russia/2000/000120-mir1.htm.

Dahlstrom, Eric. Video footage of Dennis Tito's launch tour in Baikonur, 26–28 April 2001.

Da Vinci Project. Web site. http://www.davinciproject.com.

Diamandis, Peter. "From Space to Energy: Changing the World for Good." MIT Presentation (video clip from MIT World), 27 October 2005.

————. "Questions and Answers Briefing." Supplied by X PRIZE Foundation, January 2008.

————. TEDTalks (online video). 4 September 2008. http://blog.ted.com/2008/09/taking_the_next.php.

Ehricke, K. A. "Space Tourism." Paper presented at the annual meeting of the American Astronomical Society, Dallas TX, 1967.

FAA *Environmental Assessment for the Blue Origin West Texas Commercial Launch Site*. DOT, FAA, CST. Washington DC, August 2006.

Gold, Mike, et al. Session V, "Emerging and Evolving Markets." Twelfth Annual FAA Commercial Transportation Conference, Arlington VA, 6 February 2009.

House of Travel. Web site. http://www.houseoftravel.co.nz.

Hunter, Max. Web site. http://www.maxwellhunter.com.

Kelly Aerospace. Web site. http://www.kellyaerospace.com.

Ladwig, Alan. "Space Flight Participant Program: Taking the Teacher and Classroom into Space." Address at the Thirty-Sixth Congress of the International Astronautical Federation, Stockholm, Sweden, 7–12 October 1985.

Landeene, Steve, et al. Concurrent Session I, "What Makes a Spaceport Successful?" Twelfth Annual FAA Commercial Transportation Conference, Arlington VA, 5 February 2009.

Lindskold, Anders. "Space Tourism and Its Effects on Space Commercialization." Master's thesis, International Space University, May 1999.

McMillen, Ryan Jeffrey. "Space Rapture: Extraterrestrial Millennialism and the Cultural Construction of Space Colonization." Doctoral dissertation, University of Texas at Austin, 2004.

Mojave Air and Spaceport. Web site. http://www.mojaveairport.com.

Mothership VMS Eve: Roll Out. Video. July 2008. http://www.virgingalac tic.com.

New York Press Event. Video. January 2008. http://www.virgingalactic .com.

O'Brien, Miles. "The Rutan Brothers." CNN, *Live at Daybreak,* 17 December 2003. http://transcripts.cnn.com/TRANSCRIPTS/0312/17/ lad.12.html.

Oelerich, John. "Project Private Enterprise, Inc.: A Confidential Investment Memorandum." 11 July 1980.

O'Neill, Gerard. Testimony before the Subcommittee on Space Science and Applications of the Committee on Science and Technology, U.S. House of Representatives, 23 July 1975.

Orbspace. Web site. http://www.orbspace.com.

Orphans of Apollo. Motion picture. Michael Potter (producer), Becky Neiman and Michael Potter (directors). Free Radical Productions, USA, 2008.

Palermo, Enrico. Virgin Galactic presentation. FAA AST Conference, 6 February 2009.

Peters, Fell. Letter from Gerald Bruce Levin, University of Southern California, Department of Finance and Business Economics, 17 May 1984.

Rahn, Debra. "Briefing Scheduled to Discuss Soyuz Taxi Crew Training at Johnson Space Center." NASA Headquarters Press Release 01-48, 19 March 2001.

———. "International Space Station Partnership Grants Flight Exemption for Dennis Tito." NASA Headquarters Press Release 01-83, 24 April 2001.

"Report of the Space Task Group, 1969." NASA. http://www.hq.nasa.gov/ office/pao/History/taskgrp.html.

Rocketplane Global. Web site. http://www.rocketplaneglobal.com.

Rocketship Tours. Web site. http://Rocketshiptours.com.

"Russia Hosts First Crew of International Space Station." Transcript, CNN. com, 30 September 2000.

Scaled Composites. Web site. http://www.scaled.com.

"Sea Dragon Concept. Volume I: Summary." Aerojet-General Corporation, 29 January 1963.

Simonyi, Charles. Web site. http://www.charlesinspace.com.

Space Adventures. Web site. http://www.spaceadventures.com.

Space Industrialization, Final Report. Executive Summary. Rockwell International. Washington DC: NASA, 1978.

Space Love. Web site. http://www.spacelov.org.

Spaceport America. Web site. http://www.spaceportamerica.com.

Space Tourism Market Study: Orbital Space Travel & Destinations with Suborbital Space Travel. Futron, October 2002.

SpaceX. Web site. http://www.spacex.com.

Starchaser Industries. Web site. http://www.starchaser.co.uk.

State Support for Commercial Space Activities. Washington DC: Federal Aviation Administration, 2009.

TGV Rockets. Web site. http://www.tgv-rockets.com.

Truax, Robert. "The Unholy Bible." Unpublished autobiography.

2009 U.S. Commercial Space Transportation Developments and Concepts: Vehicles, Technologies, and Spaceports. Federal Aviation Administration. Washington DC, January 2009. https://www.faa.gov/about/office_org/headquarters_offices/ast/media/Developments%20and%20Concepts%20January%202009.pdf.

Virgin Galactic. Web site. http://www.virgingalactic.com.

Whitehorn, Will. Keynote Address. Twelfth Annual FAA Commercial Transportation Conference, Arlington VA, 6 February 2009.

———. Virgin Galactic presentation. National Space Society ISDC Conference, Washington DC, 29 May 2008.

Wimmer Space. Web site. http://www.wimmerspace.com.

Witt, Stewart. Video presentation. Tenth Annual Commercial Space Transportation Conference, 2007.

XCOR Aerospace. Web site. http://www.xcor.com.

X PRIZE Foundation. Web site. http://www.xprize.org.

"X PRIZE Team Vehicle Configuration." List provided by Ken Davidian, 3 June 2003.

Zegrahm Expeditions. Web site. http://www.zeco.com.

Index

In the Outward Odyssey: A People's History of Spaceflight Series

To order or obtain more information on these or other University of Nebraska Press titles, visit www.nebraskapress.unl.edu.